ALIASES

The

for

CYBERSECURITY

 Jon Routh

Jon Routh
Aliases – The Secret Sauce for Cybersecurity

ISBN: 979-8-9893639-1-9

To my wife of nearly 50 years, and to my parents, children, and grandchildren.

CONTENTS

PREFACE

Aliases help you hide – in the clouds of planet Internet. The hidden secret to good security is hiding from those who wish to do you harm. If they can't find you, can't make contact with you, you will be safer. Hiding avoids problems that arise when your contact information becomes widely available. Cyberattacks are hard to defend against so the best thing is to avoid them by not exposing yourself to scammers. Aliases are the secret solution.

YOU HAVE PROBABLY SEEN the TV commercial that begins with the question "What's in your wallet?" A much more important question is "What's in your inbox?"

There is probably some important stuff in your mail every day, but also some spam – maybe even a lot. Probably too much of it. Spam is so common that there is a special email folder just to hold it until you examine it. Your email service provider likely filters out well over 90% of malicious spam but when there is uncertainty, some is put into your spam or junk folder for you to make the final decision about handling. Some spam "fools the bouncer" and gets in the door. It sits in your inbox waiting for you to open it.

When there is spam, there is always the possibility that it might be malicious. More importantly, there is always the the risk that you might be fooled into taking an action you might later regret. More arrives each day. Over a period of time, a failure to make the right decision is inevitable. It likely will happen when you have your guard down such as when you are tired or under stress. One mistake and you are toast.

The kind of spam you receive and how much of it is important. **Some spam is simply annoying like flys or gnats buzzing around your head on a hot muggy summer afternoon in the garden. Other spam is dangerous, more like mosquitos which carry viruses that kill more humans than any other organism on the planet.**

Spam is any unwanted communication. It can come your way via many communications "channels," such as email, phone, text messages, social media, and clickbait laden websites. We will look at this topic more deeply in a later chapter.

How well you control spam can significantly impact your security. This book reveals

a new and highly effective means of email spam control and provides step-by-step instructions for implementation.

Email spam is often considered one of the most important types of spam to control for several reasons. As one of the most widely used and pervasive forms of digital communication, email is a prime target for spammers looking to reach a large audience. Sorting through and deleting spam emails takes time and can clog up email inboxes. Many spam emails contain phishing attempts or fraudulent schemes. Spam emails can carry malicious attachments or links that, when clicked, can infect a user's device with malware or viruses. Spam is annoying and a daily frustration to deal with, and occasionally dangerous. That is why it is important to reduce or eliminate spam however you can.

My focus as a security researcher is different than most in this field. Skilled security professionals have an endless and growing list of businesses where they can get a job. There has been a significant shortage of qualified cybersecurity professionals in the United States, and the job market has been quite favorable for those with the right expertise. According to various reports and studies, the number of job openings for cybersecurity professionals in the United States is estimated to be in the hundreds of thousands. The exact number can fluctuate, reflecting the consistent demand for skilled cybersecurity specialists across industries, including government agencies, financial institutions, healthcare organizations, and technology companies.

On a global scale, the demand for cybersecurity professionals is also high. Many countries and organizations recognize cybersecurity's importance and seek skilled professionals to protect their digital infrastructure. The global job market for cybersecurity professionals is estimated to be in the millions, with numerous opportunities available worldwide. As technology advances and cyber threats evolve, the need for cybersecurity talent is expected to grow further.

What makes me different is that I am among the few who focus on improving the cyber security of individuals rather than organizations. There aren't nearly as many jobs for people in this niche area because businesses are the ones that pay, and they generally do not employ so many security people who look at the end-user side of the equation. When they do, it is their employees they are focused on. The best help they can offer customers is to be good custodians of the data they have collected and retained to keep it out of the hands of hackers.

I occupy this consumer-focused niche because I am retired, and this is one of my hobbies. I don't expect to make a fortune from doing it. If it provides a little income, I can cover the related expenses and perhaps treat myself to a nice cup of java at a local shop. The larger benefit is that I can socialize with other people through teaching and publishing. At present, I teach several courses about personal cybersecurity in the community education program at my local university.

Every good book, even non-fiction, benefits from a good story or two; this one is no exception. I have a couple of relevant stories to tell.

A Revelation

OCCASIONALLY, I SEE AN article written by a struggling journalist or blogger who has learned that using a provocative title will attract some readers. Such was the case many years ago when I read an article with a title that went something like this: "You will never tie your shoelaces the same way ever again." Or even more provocatively, it might have been written: "You have been tying your shoes wrong all your life." Like me, you have probably seen such articles covering many subjects; they are quite common.

In any case, one has to move beyond being a bit annoyed by the titles of such articles. Such titles imply that you have been doing something for a long time in a rather stupid way, or at least out of ignorance.

In the case of the referenced article, I let my curiosity rule over my ego and read the article. Indeed, I learned that there was a way to tie my shoes differently (and better) than I had been doing for decades. My shoes have stayed tied more often to this day. I learned something.

Now for the second story.

I was teaching my class at university, and a student asked me how to control spam. The student said they received hundreds of spam emails weekly and needed a solution. Frankly, I was astounded at how much spam this individual was receiving, so I asked some more questions after class and decided to dig deeper into the problem in search of a better solution that I and others had been using: having multiple email addresses and writing mail rules to filter out stuff I didn't want to see in my inbox.

A question often asked in email-related discussions is how many personal email addresses a person should have (outside of work). I have previously asked myself this question, and after some research on the topic, I created some new email addresses and began using them with my online accounts. I had one address for financial institutions, another for shopping, etc. The idea behind multiple addresses is that if one gets compromised, you don't have to change so many accounts to eliminate a spam problem when it happens. The problem with this approach is that you must check multiple email accounts, which becomes a hassle if you handle your mail on the web through a browser. This is not a concern if you have mail delivered to your local computer from various email accounts. Mail rules, or filters as they are sometimes called, examine all incoming accounts as if they were one, and I use them to move mail to folders. More on this later.

I had never deeply explored creating aliases instead of new email accounts. I overlooked using aliases because they were described as a means of addressing another problem. Most likely, you have seen advertisements or articles that suggest you create a "**burner**" account when you want to get past a paywall to get a coupon or a "free" pdf file on some topic of interest. A burner account is just an alias. You create it and use it one time and then discard it. Most burner addresses have a time limit, so they will be deleted and reused by the supplier after the time has passed. They are useful for the intended purpose but not for much else – or so I thought.

Spam control is a serious security concern. The historical way of dealing with spam was to filter it. That is how I handled it, as best I could. In agriculture, threshing is the process of separating the wheat from chaff. In email handling, filtering separates valued emails from undesirable spam.

Spamming is the first step in the process of scamming someone, most account takeovers, and much malware distribution. The more spam you get, the higher the risk of getting duped or hacked. Businesses know this, and many require employees to attend training classes to learn to recognize and avoid phishing emails which cost businesses dearly when successful. News headlines regularly mention organizations severely affected by ransomware infections that begin with Business Email Compromise, or BEC. Malicious spam is a problem, and the easiest entry point to any business is the compromise of user accounts.

A Pivot

MANY STARTUP BUSINESSES LAUNCH with an idea for a new product or service that sounds good enough that it attracts some investor financing. Sometimes, things don't go as planned, however.

A pivot is a strategic decision made by a company when it realizes that its newly launched product or service, initially designed to address a specific need, possesses untapped potential in a different area. It involves recognizing market feedback, customer insights, or unexpected usage patterns that reveal distinct needs or opportunities beyond the original intention.

When a pivot occurs, a company proactively adapts its approach to capitalize on the discovered potential. This could involve reshaping the product's features, repositioning the marketing strategy, or redirecting the entire business model. By doing so, the company seizes the chance to cater to unanticipated market demand or underserved customer segments.

The pivoting process may involve conducting further research and analysis to validate the identified need's viability and refine the product or service accordingly.

In this scenario just described, I was not the business but rather the customer who recognized the potential value of a service (aliases) when applied to solve a different problem: malicious spam.

An Aha Moment

An "aha moment" is a powerful instant when everything clicks into place, and a person experiences a profound revelation or understanding. It often arises after contemplation, analysis, or exploration of a particular subject or challenge.

I had one of those. I didn't need multiple personal email addresses; I needed only two—one for personal and one for personal business. What I needed instead was **unlimited permanent aliases** - ok, just a hundred or so would be fine. It would be better to stop emails from being sent to me in the first place rather than trying to filter out the good from the bad after it was already in my inbox.

I am not the only person who has thought of this approach and surely not the first to try it out. However, I am pretty certain I am the first to write a book about it – to tell others what a difference it can make.

Like many a researcher, I tested this hypothesis first on myself. I needed to understand the difficulty of implementation and the amount of effort involved. I also needed to measure the results and identify any unintended consequences.

I already mentioned that I have begun teaching my methods of email spam control to adults in community education. Now I am turning my sights to working with local public schools in curriculum development. Those of us who are adults have already established communications with numerous businesses using real email addresses. To regain control over spam, we must take recovery actions ... and those actions take time to implement the required changes.

Young people just starting out in life as they exit high school have the opportunity do things right from the beginning. Parents of young teens should take to heart what I present in this book and guide their children to hide their email addresses, thereby avoiding exposure to the scams and infections that we "older ones" currently experience in our daily lives.

Let's get started!

Chapter One

SPAM - THE HIDDEN DANGER IN YOUR INBOX

BEFORE TAKING A DEEP dive into the world of email, let's look at a diagram of typical mail flow across the Internet. The purpose of the diagram is to show how email makes its way from sender to recipient. Figure 1-1, depicts the flow of email, much of which is spam. Figure 1-2, which follows, is a numbered list of descriptions of the important stages of mail flow.

Sources and Flow of Email

In the diagram, various parts of the Internet (the "Cloud") are shown.

- The **Open Web** is customer-facing and is indexed by search engines like Google.

- The **Deep Web** is also customer-facing but is gat-

ed. You must have credentials to log in to a website to see the contents of the deep web. This
is where users with accounts transact business.
Private Networks, also part of the deep web, are
the back-end systems that handle the day-to-day
computing tasks that run a business. Only employees and some business partners can access
an organization's internal networks.

- The **Dark Web** is where many organizations do
business, including criminal hacking groups. You
must have a special browser (called TOR) to
reach websites on the Dark Web. Hacker groups
operate marketplaces where among other things,
data are bought and sold.

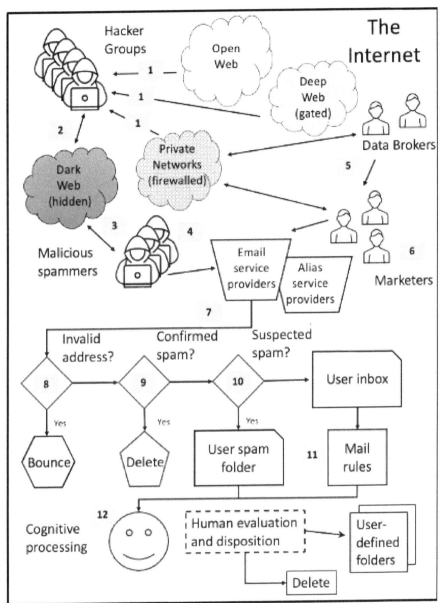

FIGURE 1-1. Mail flow diagram

1	The arrows indicate that the hacker groups steal data from the first three networks and store their information in the fourth.
2	Hacker groups offer data for sale, including email address distribution lists. They also develop and market "exploit" (attack) kits for others to use.
3	Malicious spammers compose emails and send them using address list obtained from the dark web.
4	Malicious spam is sent to Alias Service Providers (ASPs) or directly to Email Service Providers (ESPs) for delivery to a user account.
5	Data brokers sell address lists to marketers
6	Organizations send emails to customers and prospects using homegrown or purchased mailing lists. These emails may be spammy but are harmless.
7	Emails are processed by ASPs and ESPs as they are received.
8	When an address doesn't (or no longer) exists, the email is returned to the sender with an error. This is called a "hard bounce."
9	When filters applied by an ESP confirm an email is spam, it is deleted.
10	Mail that passes spam filters is put into a user inbox or spam folder.
11	Mail is passed through any user-created mail filters and perhaps moved to other folders.
12	The owner of the email account makes decisions regarding final disposition of email received.

FIGURE 1-2. Mail flow diagram descriptions

In the following chapters, we will discuss email aliases and Alias Service Providers, or ASPs, (mentioned in item 4) in more detail.

ASPs act as a control point you can use to prevent email from being sent onward to an ESP. ESPs maintain mailbox storage, perform filtering and enable user access. ASPs serve in a role similar to a bouncer at a nightclub or a doorkeeper in a large residential building. They are there to keep unauthorized people out.

If you can stop malicious spam from arriving in your inbox, you will be less reliant upon additional defenses you may or may not have put in place.

This is important because not enough people maintain sufficient awareness of social engineering attacks or sustain the vigilance required to avoid making a mistake such as clicking on a link or attachment before verifying it is safe. Additionally, many people are ill-equipped to implement the many technically complex recommendations made by security experts to protect personal networking and computing devices. As a result, too many fall prey to such attacks.

In the following chapters, I will show you how to implement this new alias based approach to improved personal cyber hygiene.

The Scourge of Spam

SPAM REACHES EVERY CORNER of the world, making up almost 70% of email traffic. Email service providers go out of their way to protect users from this problem with an arsenal of sophisticated filters. Still, some spam always makes its way through and ends up in your inbox or spam folder.

People who grew up in a big city know, among other things, that nothing good comes from a stranger approaching you and saying, "Can I ask you a question?" You'll most likely be locked into a long conversation that ends with a plea for money. This is why people living in dense urban environments walk as if on a mission and avoid eye contact with others. The phrase "street-smart" is frequently used to describe these individuals.

Email is an "approach" from one party to another to start (or continue) a digital conversation. Therefore, to avoid problems, we all need to be "scam-smart" because we no longer live in the safe village of Camelot. We now live in the crime-ridden megacity called Spamalot. Most of us suffer an ever-rising tsunami of spam in our email inboxes and spam folders and wonder what can we do about it.

Until now, no one has proposed a solution that doesn't require knowledge of writing complex mail rules. In this book I will show you how to implement strong digital pest control that doesn't require programming skills and that is extremely effective.

Large and small organizations are regularly attacked and often breached, exposing our data. This trove of information, which often includes email addresses, is used by bad actors in many nefarious ways.

Any student of military history knows how superior weaponry gives attackers great advantages in conquering others. Wooden shields cannot stop bullets or bombs. Humans are highly vulnerable to social engineering techniques employed in malicious spam. Also, many of us practice poor cyber hygiene for various reasons.

For example, the latest statistics show that at most twenty-five (25) percent of consumers use a password management application (app) to store the credentials required to access online accounts. Anyone who tries to rely on memory alone to recall passwords will take shortcuts – using weak passwords and reusing passwords across numerous accounts. If you can remember your password, it is surely a very weak one. Modern AI tools can crack a weak password in seconds.

New Generative AI tools enable hackers to write perfect phishing emails in any major language, mimicking the style and format of legitimate communication from any large organization and suffering no misspelled words or grammatical mistakes. These tools can also be employed to write shape-shifting malicious code that can avoid detection by spam filters.

Native cultures occupying the Americas could not successfully defend themselves against Europeans arriving with guns, germs, and horses. They got wiped out. Digital citizens need a new means of defense against scammers coming with superior weaponry. It is time to change the rules to win the war.

Assigning a unique alias to each online account and replacing it when compromised is like stopping someone from approaching you on the street. If a bad actor doesn't have a valid email address, they cannot approach you with a scam, infect your computer, steal your identity, or do any number of terrible things. Maintaining control of your CONTACTABILITY is the new means of defense. It is a game changer.

Time is running out. As we enter the age of Artificial Intelligence (AI), **bad actors WILL up their** game by using these astoundingly powerful new technologies to overwhelm us with much higher success rates.

This book describes how we got into this unsustainable situation and how we can regain control. The tools are now freely or inexpensively available. The method is what has remained unclear. With a way forward identified, it is now possible to regain control of our digital lives.

Types of Spam

HERE IS A LIST of common spam categorized by communications channel/platform with brief descriptions:

- **Email Spam:** This is one of the most well-known forms of spam. Email spam refers to unsolicited bulk emails sent to many recipients. These messages often contain harmless advertisements but also may contain hidden dangers such as scams, malicious links, or malicious attachments. Malicious emails have one of two subtypes. The first, a phishing email, aims to trick recipients into revealing sensitive information. The second subtype is designed to trick recipients into downloading malware onto their devices, allowing attackers to gain unauthorized access or steal sensitive information.

- **Social Media Spam:** Social media platforms are also used by spammers. Social media spam can be unwanted messages, fake accounts, deceptive links, or irrelevant comments intended to promote products or manipulate users.

- **Voice spam:** All telephone numbers are susceptible to calls from unknown or spoofed numbers, making it difficult to trace or block them effectively. They are the number one complaint by consumers to the Federal Communications Commission. Voice spam refers to the unsolicited and unwanted mass distribution of voice messages or robocalls to many individuals. Like email or text message spam, voice spam aims to deliver promotional messages, scams, or fraudulent schemes through automated phone calls.

- **Instant Messaging Spam:** Like email spam, instant messaging platforms can also be plagued by spam. Users may receive unsolicited messages containing advertisements, phishing attempts, or malicious links through instant messaging services.

- **SMS Spam:** SMS spam involves sending unsolicited text messages to mobile phones.

These messages typically contain advertisements, scams, or phishing attempts. SMS spam
can be intrusive and may lead to unwanted
charges for the recipient.

- **Comment Spam:** Comment sections on websites, blogs, or forums are often targeted by
spammers. Comment spam involves posting irrelevant or promotional messages to drive traffic
to specific websites or spreading malicious links.

- **Forum Spam:** Online forums and discussion
boards are common targets for spammers. Forum spam includes posting irrelevant or promotional content, often with links to external websites. This type of spam disrupts conversations
and may contain malicious links.

Email Spam

IN THIS BOOK, WE are going to concentrate on eliminating bulk spam emails, whether malicious or benign.

Email is the primary method of communication between organizations and between organizations and
individual consumers because it is a ubiquitous service
provided by the Internet. Statistically speaking, email

is the first step in nearly all successful data breaches, malware infections, and scams.

Stopping spam email will dramatically improve your security, boost your productivity, and give you peace of mind.

Let's begin by defining what we mean when discussing bulk email spam. Here are some categories of email spam:

- <u>Advertising</u>: This is the safe kind of spam. Its purpose is usually to promote products or services. Advertisements can range from legitimate marketing campaigns to deceptive or illegal offers. They often include persuasive language, flashy graphics, or enticing discounts to lure recipients into purchasing.

- <u>Scams and fraud</u>: Scam emails come in various forms and aim to defraud recipients by promising financial gain or exploiting their vulnerabilities. Common scam emails include lottery, inheritance, romance, or job opportunity scams. The goal is usually to extract money or personal information from unsuspecting victims.

- <u>Phishing</u>: Phishing emails are a particular type of fraud designed to trick recipients into divulging sensitive information such as usernames, passwords, or credit card details. These emails typically appear to be from reputable organizations, and they often use urgency or fear tactics to trick recipients into clicking on malicious links or providing personal information. An example of a malicious link impersonating another website. For example, a link included in an email may be labeled as "Chase Bank," but the target link (which is hidden until revealed by the recipient) might be something like "chasebank.com," whereas the correct domain would be "chase.com." The unwary user may be tricked by this and attempt to log in to the bogus website using their bank credentials. This trick has a name; it is called spoofing.

- **Malware distribution**: Emails may contain malicious attachments or links that, when clicked or opened, install malware on the recipient's device. Malware can take various forms, such as viruses, ransomware, or spyware. These emails often masquerade as important messages, invoices, or delivery notifications to entice recipients into opening the attachment or clicking the link.

- **Adult content solicitation**: This spam email contains explicit or adult content or promotes dating or adult websites. These emails often use attention-grabbing subject lines or provocative images to entice recipients to open and respond to them.

Sending spam is illegal in many jurisdictions, and legitimate businesses and organizations typically follow anti-spam regulations, such as obtaining proper consent from recipients before sending commercial emails. However, spammers who engage in illegal activities often operate outside the law and ethical boundaries. It is crucial to remain cautious about sharing your email address online and protect your personal information.

Frequent Spammers

IMAGINE YOU ARE A **marketing professional** and want to send emails to a broader audience than your current customer base, so you need to obtain a mailing list from a third party. Here is a list (in no particular order) of some of the criteria you might use to choose a provider:

- **Relevance:** The email list should align with your target audience and the purpose of your campaign. For example, a list of technology enthusiasts or professionals would be more suitable if you're promoting a technology product.

- **Quality and Accuracy:** A high-quality list would have accurate and verified email addresses. A marketer would want to ensure the list is regularly updated and the emails are active and valid. An outdated or inaccurate list would result in wasted effort.

- **Opt-in Subscribers:** Opt-in email lists, where individuals have given explicit consent to receive emails from third parties, ensures compliance with anti-spam regulations and increase the likelihood of engagement.

- **Demographic Information:** Depending on a campaign's objectives, a marketer might prioritize email lists with relevant demographic details such as age, location, industry, or interests. This allows for more targeted and personalized messaging.

- **List Size:** The size of the list may be a consideration, as it determines the potential reach of a campaign. Quality is prioritized over quantity. An engaged and smaller list may yield better results than a large, unengaged one.

- **List Source:** Understanding the source and origin of the list is crucial. Preferably, a list was obtained ethically and legally, from trusted sources or reputable vendors.

- **Engagement Metrics:** Assessing subscribers' engagement level on the email list can be valuable. A list with a history of high engagement (high open rates, click-through rates, and conversion rates) indicates a more active and receptive audience.

- **Niche Specificity:** If you're promoting a niche product or service, a list specific to that industry or interest group could be more valuable.

- **Geographical Targeting:** If a campaign is region-specific, consider email lists that target a particular geographic location or language group.

- **Opt-Out and Unsubscribe Rates:** Evaluating the opt-out and unsubscribe rates can provide insights into the quality of the list. A lower opt-out rate indicates that subscribers find the content valuable, while a high rate may suggest low engagement or irrelevant targeting.

- **List Maintenance and Hygiene:** A well-maintained list is regularly processed to remove inactive or non-responsive email addresses. A clean list reduces bounce rates and improves deliverability.

- **Reputation and Trustworthiness:** Consider the reputation of the email list source. Reputation matters when it comes to building trust with your audience. A reputable and trustworthy list source will likely provide valid and reliable email addresses.

- **Cost and Budget:** Assess the cost of renting or acquiring the email list. While it's crucial to prioritize quality, you should also consider whether the list aligns with your budget and offers a reasonable return on investment.

Now, Imagine you are a hacker who targets consumers, not organizations, and you want to send emails for a phishing campaign. You could rent a list from a legitimate source, but among other reasons, the cost per thousand is very high, so instead, you turn to the dark web in search of sources of email distribution lists.

Would the marketer and hacker have different criteria for selecting one distribution list over another beyond price? **It turns out that both marketers and hackers have very similar needs for the most part.** There are, however, some significant differences.

Here is a list of email distribution list selection criteria that would be used for nefarious purposes:

- **List Size:** A larger list provides for wider reach and potential impact.

- **List quality:** A clean list reduces bounce rates and improves open rates. Spammers don't want to send emails to inactive accounts. They expect lists to be cleaned of these. Otherwise, they are paying for addresses that don't have the potential to yield results.

- **Geographic targeting:** see the description previously provided.

- **Demographic Information:** see the description previously provided.

- **Vulnerability:** Targeting lists that include individuals with known vulnerabilities or weak security measures could be advantageous.

- **Social Engineering potential:** Evaluating the potential for social engineering and manipulation based on available data and personal information within the list.

- **Security Measures:** Assessing the security measures that certain users might have, such as two-factor authentication, to target those less likely to have implemented these measures.

- **Reputation:** Evaluating the reputation of the list source from a malicious standpoint, such as identifying sources known for distributing compromised or hacked data.

- **Cost and Anonymity:** Considering the price and level of anonymity associated with renting or acquiring the list to minimize the risk of exposure.

As you can see, there are important differences between these two lists of criteria. Those who maintain lists for rent to others for nefarious use will create very different categories than those who rent lists to marketing professionals for legal use.

Targeting

SCAMMERS TARGET VARIOUS CATEGORIES of consumers based on their vulnerability and potential for financial gain. While it is important to note that anyone can become a victim of scams, certain groups tend to be more frequently targeted due to specific characteristics or circumstances. Here are some categories that bulk email scammers commonly target:

1. **Former scam victims:** Scammers often target individuals who have previously fallen victim to scams. Using "Sucker lists," they prey on previous victims or databases of individuals who have shown vulnerability to carry out follow-up scams, such as recovery or "reload" scams.

2. **Elderly individuals:** Scammers often target senior citizens due to social isolation, lack of familiarity with technology, and potential cognitive decline. They may exploit their trust or use scare tactics to deceive them into providing personal information or money.

3. **Financially distressed individuals:** Scammers often target people experiencing financial difficulties or seeking quick solutions. Desperation or a desire for easy money can make them more susceptible to fraudulent schemes promising quick profits, loan scams, or debt relief scams.

4. **Online shoppers:** With the increasing popular-

ity of online shopping, scammers target consumers who make purchases on e-commerce platforms. They may use fake websites, counterfeit products, phishing emails, or fake customer service representatives to steal personal information or payment details.

5. **Job seekers:** Scammers prey on individuals seeking employment by offering fraudulent job opportunities. They may request payment for fake background checks and training materials or promise high-paying positions to extract money or personal information from job seekers.

6. **Immigrants and non-native language (e.g . English) speakers:** Scammers often target immigrants or individuals lacking local language fluency. They may exploit their unfamiliarity with local customs, laws, or institutions to deceive them through immigration scams, fake legal services, or fraudulent job offers.

7. **Social media users:** Scammers take advantage of the widespread use of social media platforms. They may use social engineering techniques to trick users into sharing personal information or clicking on malicious links, leading to identity theft, financial fraud, or malware infection.

Let's examine a bit more closely a particular demographic from the list above. Here are some additional characteristics of older generations that make them a prime target of spammers:

1. **Limited technical knowledge / skills:** Many seniors are less familiar or comfortable with technology and the internet than younger generations. This lack of technical expertise and skill can make them more vulnerable to scams and phishing attempts. Spammers take advantage of this knowledge gap and exploit seniors' limited understanding of online security.

2. **Experiencing age-related cognitive decline:** As individuals age, some may experience cognitive decline, including issues with memory, decision-making, and critical thinking. Cognitive decline can impair a person's ability to detect red flags, assess risks, and recall previous warnings or education about online security. This puts them at a higher risk of falling for scams or being tricked into sharing sensitive information. They may also find responding appropriately to suspicious or fraudulent communications more challenging. Spammers exploit this vulnerability by employing urgency, fear, or manipulation tactics to convince seniors to divulge personal information or engage in harmful actions.

3. **Having a trusting nature:** Certain generational groups tend to be more trusting than others, which can make them susceptible to scams. They may be more inclined to believe fraudulent

emails or phone calls and provide personal information or financial details without verifying the source's legitimacy.

4. **Possessing wealth:** Seniors are often perceived as having accumulated savings and assets, making them an attractive target for scammers who aim to exploit their financial resources. These scammers may offer investment opportunities, fraudulent products or services, or impersonate family members needing financial assistance.

5. **Experiencing Isolation and/or loneliness:** Some seniors may experience social isolation, leading to a desire for human connection and interaction. Spammers exploit this vulnerability by sending phishing emails or messages that appear to be from a friendly or caring source, taking advantage of the seniors' need for companionship.

6. **Experiencing medical and/or healthcare issues:** Seniors are more likely to be concerned about their health and well-being, making them susceptible to scams related to medical products, treatments, or health insurance. Spammers may send misleading or false information about miracle cures or health-related services to exploit their concerns.

Sources of Email Addresses Used by Malicious Spammers

Suppliers of email distribution lists to hackers are criminal enterprises and do business on the dark web. If your email address gets on any of these lists, the only way to stop receiving malicious spam is to change your email address. As long as the email address remains valid, it can only be dealt with by filtering. Let's now look at how personal email addresses end up in criminal hands in the first place. Spammers acquire email address lists through various means, and here are some common methods:

1. **Data Breaches:** Hackers illegally enter private networks and exfiltrate (copy) user information, including email addresses. The first and largest source of consumer email addresses comes from these breaches. To make the most from what they have stolen, hackers sell or rent the data on to other malicious agents once other attempts at value extraction have taken place. It is a domino effect; a company gets hacked, customer data is stolen, and the data is sold to others.

2. **Harvesting:** Spammers use automated tools to scan websites, forums, and social media platforms for email addresses. These tools extract email addresses from public sources or by crawling through web pages, collecting any visible email addresses they come across.

3. **Malware and Botnets:** Malware infections on users' computers can enable spammers to access their email contact lists. Botnets, networks of infected computers controlled by spammers, can also collect email addresses from compromised devices.

4. **Purchased Lists:** Some spammers buy email address lists from third-party sources. These lists may be acquired legally or illegally and often contain addresses collected without the individual's consent. Data brokers — "people finder sites" — are legitimate companies that collect and sell information about you. While their main clients are advertisers, scammers can buy personal contact information lists to power their schemes.

5. **Subscription Lists:** Spammers may scrape or purchase subscription lists from legitimate companies, disregarding proper consent and abusing the acquired email addresses for spamming purposes.

6. **Shared Lists:** Spammers share lists of active email addresses based on your behavior. If you click on a link or even open a spam email, it tells them you're a likely good target.

Are You Exposed And Is Spam a Problem For You?

THE REAL QUESTIONS EACH of us need to ask ourselves:

1. Are we are exposed by having contact information available to scammers on the dark web

2. How much malicious spam are we getting in our inbox(es) and spam folder(s).

There is a web-based tool will let you enter your email address or phone number and see if either has been found on the dark web. The tool is free of charge and will tell you if your email address OR phone number has been exposed in one or more data breaches. The tool will also identify which breach exposed your email address, the date it was reported, how many users were affected, and what data, in addition to your email address, was likely exposed.

The tool is free to use without creating an account at this URL: `HaveIBeenPwned.com`. The QR code printed on the back cover of this book, when scanned with your phone, will take you to this website.

The database searched by this website includes information about large (> 1,000 user accounts) <u>reported breaches</u>. **US citizens should note that the data is quite incomplete** because there are no laws at the federal level that require organizations to report data breaches. Nonetheless, the tool provides important insight into your current level of exposure.

If you learn through use of this tool that your email address (or an alias) has been exposed one or more times, then you should strongly consider taking actions recommended in this book. Learning that your address has been exposed in just one data breach indicates that you have been put at risk. Learning that your email address has been found in more than one breach doesn't necessarily indicate higher risk. Instead, it indicates how many times you could have eliminated the risk by disabling the alias and assigning a new one. Only had you been using unique aliases could you perform this action easily and quickly.

Identifying Potentially Malicious Emails

There are several tell-tale signs that an email is potentially dangerous. It is crucial to consider multiple indicators collectively and use your judgment to assess the overall legitimacy and safety of an email before you do anything other than delete it.

Exercise caution if you encounter an email exhibiting one or more of the indicators in the list below. Avoid clicking on links, downloading attachments, or providing personal information unless you are confident about the email's legitimacy. If in doubt, delete the email. If you have concerns and wish to investigate further before taking any action, it is wise to contact the purported sender through a separate and verified communication channel to verify the email's authenticity.

Here is a list of attributes of an email that should raise suspicion:

1. **Suspicious sender address:** The sender's email address must be checked carefully. Malicious emails often use fake or spoofed addresses that mimic legitimate organizations or individuals. Look for misspellings, unusual domain names (see next section), or variations in the sender's address.

2. **Unusual or unexpected email content:** Malicious emails may contain unusual or shocking content. This can include requests for sensitive personal or financial information, offers that seem too good to be true, or urgent demands for immediate action.

3. **Poor appearance:** Many malicious emails exhibit poor spelling, grammar, and formatting errors.

4. **Urgent or emotion triggering language:** Malicious emails often use urgency or threats to manipulate recipients into taking immediate action.

5. **Suspicious attachments or links:** Emails containing suspicious attachments or links to unfamiliar websites.

6. **Generic greetings or lack of personalization:**

The email uses no greeting or a generic one like "Dear Customer."

7. <u>Lack of a "Reply To" address</u>: Many businesses include a "Reply To" address to route replies to the correct department for proper handling.

8. <u>Mismatched URLs or suspicious domains</u>: The link text does not match the target URL, or the domain looks suspicious or unfamiliar.

9. <u>Unexpected file downloads</u>: Emails that automatically trigger file downloads when opened or prompt you to download files from unfamiliar sources should raise suspicion.

Some have used email rules to deal with spam not caught by Email Service Providers to identify suspicious emails automatically. Writing and maintaining such rules can be a technically challenging task. This approach is not a viable solution for most people.

Abused Internet Domains

INTERNET DOMAINS ARE SOME of the most highly abused tools threat actors use to manipulate victims and execute phishing attacks. PhishLabs Quarterly Threat Trends & Intelligence report issued in September of 2021 tells the story of abuse. Bad actors register Legacy Generic (gTLD), Country Code (ccTLD), and newly added Top-level domains, HTTPS, and free security certificates to target victims.

In the United States, most legitimate business email comes from domains that end with the.COM top-level domain (TLD), as mentioned in the PhishLabs report. In many smaller countries, some mail comes from domains that end with what is referred to as a County Code TLD (ccTLD) such as UK for the United Kingdom, FR for France, and so on. Finally, some very small countries make money by using their ccTLD for free or at a low cost for registration. There are also many legacy (gTLD) and newly approved TLDs.

Figure 1-3 below shows the 10 Top Level Domains (TLDs) that were the most highly abused by threat actors to manipulate victims and execute phishing attacks.

TLD (Top Level Domain)	Type	% Phishing sites
.COM	Legacy gTLD	39.7%
.ORG	Legacy gTLD	5.9%
.CA	ccTLD	4.1%
.IO	ccTLD	3.7%
.NET	Legacy gTLD	3.2%
.MX	ccTLD	2.8%
.CO	ccTLD	2.4%
.UZ	ccTLD	2.4%
.MONSTER	New gTLD	2.2%
.AE	ccTLD	2.1%
All other	varies	31.5%

FIGURE 1-3. Abuse of top-level domains. Phish Labs 9/21

Nearly 40 percent of phishing websites use a COM top-level domain, which isn't surprising. The report also states that historically, five country code TLDs (ccTLDs) are often on the abuse list because bad actors can register them through a known free domain provider: The five TLDs that were the worst offenders at the time of the report were ML, TK, GA, CF, and GQ.

It pays to be suspicious of domains that use an uncommon TLD. In a topic presented later, I provide an example of how to write an email rule to identify unusual TLDs.

Many people still think the website is safe when they see HTTPS (vs. HTTP) at the beginning of a link (URL). The truth is that the "S" at the end means that the communication that will be established between a browser and the website will be secure (encrypted) but does not indicate that the website is safe. The reality is that over eighty percent (82%) of phishing websites employ HTTPS to trick uninformed users.

In the next chapter, we will look at some current spam control methods and reveal a new approach.

Chapter Two

METHODS OF EMAIL SPAM CONTROL

THE BEST POSSIBLE OUTCOME of spam control measures is eliminating the nasty, dangerous stuff and limiting the amount of safe but annoying stuff.

Email services are available through many sources. Some people use the service their Internet Service Provider (ISP) offers but a better option is to obtain email services through an alternative source such as a large company like Alphabet (Google Gmail), Microsoft, or Apple. All Email Service Providers (ESPs) provide mail filtering (spam control). The quality varies greatly, so if you want better filtering, you are better off <u>not</u> using the services of your ISP because they have little incentive to invest in filtering technologies than companies with nationwide or worldwide coverage.

Choosing An Email Service Provider

Webmail is an online-based email service enabling users to access their inboxes and mail folders without additional software. Sign up for an email address with an ESP, then to do email just navigate to that email provider's website to log in to your account through your browser. You can do so from any device, anywhere.

The most popular free email service providers are actively used by millions of people worldwide. Alphabets' Gmail has the largest user base, according to 2022 data, with a 49% share, followed by Apple (iCloud) with 28% and Microsoft (Outlook) with 13%.

Choosing the right provider is important because there are large differences in the quality of spam filtering offered. You will be better served by choosing email services offered by one of the three mentioned global providers than by an offering available through a local ISP. Better filtering means less spam which results in reduced risk.

There often needs to be more clarity when contrasting webmail and email clients. Email clients provide a special device level app that retains as much (if not more) of the same functionality as webmail. Those programs, such as Microsoft Outlook, Apple Mail, or Google Gmail, use similar names to their webmail counterparts, adding to the confusion. These apps connect to your email server behind the scenes and download the information to your devices for handling.

Webmail interfaces are popular for their portability, security, and compatibility. Still, desktop email client users prefer their software because they may wish to manage multiple email accounts in a single interface or need a more vigorous backup solution.

Suppose you create multiple online email accounts, one for personal correspondence and another for private business. In that case, it makes sense to create both using a single ESP if you intend to create mail folders and write rules/filters on the server side. This is because you will not have to learn more than one user interface. If you create rules and folders on a local system, using more than one ESP makes little difference since you can access more than one service through a single client-side interface. In this case, the determining factor in choosing an ESP has more to do with the strength of spam filters offered and privacy-related aspects.

Spam filtering methods employed by ESPs

Spam filters are designed to detect unsolicited, unwanted, and virus-infected emails and prevent them from reaching email inboxes. Spam filters can be applied to incoming emails (entering the network) and outgoing emails (leaving the network).

ESPs employ several approaches to controlling spam at the server level before it reaches a user inbox. Some of these are:

- Block lists of previously identified IP addresses and emails

- Allow lists that specify domains that will bypass an ESP spam filter.

- Rule-based detection identifies common patterns associated with known threats. This includes checking for spelling and grammatical errors and reviewing a list of trigger words heavily featured in known spam emails.

- Signature-based detection to identify previously flagged malicious attachments or links within emails, for example.

- Bayesian analysis which deduces from various attributes the probability that an email is spam.

- Databases containing spam for comparison.

- URL verification.

- Artificial intelligence technologies such as machine learning algorithms.

ESP spam filtering effectiveness

The Gmail team has claimed to have conducted some research on their system that found that 99% of legitimate mail got through while 99.9% of spam was blocked. Those are impressive numbers if true. On the other hand, since an estimated 110 billion spam emails are reportedly sent daily, far too many still make it through and put users at risk.

Email filters are designed to catch and block a wide range of fraudulent and scam emails, but some types of emails still manage to evade detection. Here are a few examples of scams that could potentially bypass email filters:

1. **Social engineering tactics:** Some scams rely on psychological manipulation rather than malicious content. These emails may not contain any obvious red flags, making it harder for filters to identify them as fraudulent. Such attacks may not contain any links or attachments but instead request a reply by email or by phone. Some common examples include romance scams and investment scams.

2. **Email spoofing:** Scammers can forge the sender's address to make it appear as if the email is coming from a trusted source, like a reputable company or a friend. Advanced email filters may detect spoofed emails, but some still slip through.

3. **Image-based spam:** Instead of using text, scammers embed their messages within images to avoid text-based filters. While some email filters can analyze image content, this technique can still be effective at bypassing detection.

4. **Encrypted or password-protected attachments:** Scammers might encrypt or password-protect malicious attachments to evade detection by filters that cannot analyze the content inside en-

crypted files.

5. **Polymorphic malware:** Polymorphic malware constantly changes its code, making it difficult for signature-based filters to recognize it. This allows scammers to deliver malware-laden attachments undetected. AI technologies are making the creation of polymorphic malware an easier task for hackers.

6. **Zero-day attacks:** These are newly discovered software vulnerabilities or exploits that have not yet been patched. Scammers can use zero-day attacks to deliver malicious content without being detected by traditional filters. *Consumers are unlikely to be affected by these types of attacks because they are reserved for use on high value targets like businesses.*

7. **Spear-phishing emails:** These are highly targeted scams that are tailored to specific individuals or organizations. Since they are customized, they may not trigger typical spam detection rules. Consumers are unlikely to receive such emails.

8. **Legitimate but unwanted emails:** Some emails may be technically legitimate (i.e., not containing malware or scams) but are still considered unwanted by the recipient. These could be advertisements, newsletters, or promotional emails that the user never subscribed to.

Scammers are constantly evolving their tactics to bypass filters, which is why it's essential for users to stay vigilant and be cautious when dealing with unsolicited or suspicious emails.

A More Effective Method Than Filtering

MOTHER NATURE HAS ENDOWED some species with the ability to ward off predators with camouflage. The chameleon can even adjust its skin color to match its surroundings. We can do something similar.

Rather than rely on automated filtering or repeated examination of incoming emails, **this book describes a much more effective approach: keeping your email address cloaked or hidden by using aliases.**

First, we will look at eliminating malicious email, and in the final section, we will look at reducing or eliminating the annoying but benign spam sent out by advertisers.

One person may consider a frequent notification of products on sale at a favorite store as desirable information, while another will call it spam. In any case, email sent to you by organizations you do business with is best handled by selecting your communications preferences in the profile section of your online accounts. This type of spam control will be discussed in a later chapter. For now, we will focus on eliminating malicious spam from your inbox.

Some emails are placed in our spam or junk folder when the analysis of emails performed by an ESP results in some uncertainty. Some spam gets through, and some of it is malicious. Beyond the security aspects we have been discussing, there are also some privacy issues to consider when selecting an ESP.

The larger ESPs such as Google, Microsoft, and Yahoo have the incentive to offer mail handling services for "free" in exchange for having access to our communications, and this data is of high value to marketers.

Apple has a more privacy-focused business model because it makes money from selling hardware and services rather than relying on advertising for profit. Apple has recently brought to market a trio of privacy-focused services for iCloud email users that pay for iCloud Plus, ranging from $1 to $10 per month. While most of the buzz about these features concerns privacy, the potential to use these new features for improved security is a game changer. I will describe how to leverage these new features in this book. To take advantage of these services (from Apple), you first must have an iCloud account with Apple.

Choosing Apple for Alias Handling

ICLOUD IS APPLE'S CLOUD storage and syncing service that allows you to store and access your data across multiple devices. However, to use the account on a Windows computer, you must also install **iCloud for Windows** to access your photos, videos, mail, calendar, files, and other important information on your Windows PC. The app is available from the Microsoft store.

There may be more attractive options than this approach, however. If you do not own any Apple devices, you may prefer to choose another alias service. Several such options will be discussed in a later chapter. Regardless of service provider, the benefits are the same, and the implementation process is similar to that described here.

We all know an email address, but only some are as familiar with an email alias. Aliases are commonly used by businesses but rarely by individuals. Organizations must route various incoming emails to an appropriate department for proper and efficient handling. People move in and out of departments, so routing needs to be adjusted to add or delete people from teams. The Table in Figure 2-1 below demonstrates how mail sent to generic addresses is routed internally based on how a business is structured and staffed.

Alias name	Alias Address	Forwards to
Info Alias	info@company.com	Public Relations team
Support Alias	support@company.com	Customer Support team
Sales Alias	sales@company.com	Sales team
Admin Alias	admin@company.com	Website technical support team
Marketing Alias	marketing@company.com	Marketing department
No Reply Alias	no-reply@company.com	Routes replies to an unmonitored inbox (by convention)

FIGURE 2-1. Business use of aliases

Aliases look similar in format (xxxx@yyy.tld) to real email addresses but are handled differently. Both types of services have been available to consumers for over two decades. The graphic in Figure 2-2 below demonstrates how an alias works for personal use. In the example, each of the three websites is given a different email alias instead of a real email address. The alias service provider receives emails and routes them to a single real email inbox. One of the important things about an alias service is that if you reply to an email sent to your alias, the reply will be sent back through the alias handling service first rather than directly to the original sender. This is because a "reverse alias" process needs to be performed. This is required to preserve your privacy by not revealing your real email address.

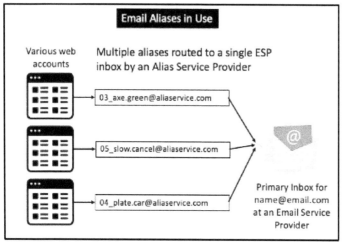

FIGURE 2-2.How personal email aliases work

A Bit of Email History

THE FIRST <u>FREE</u> EMAIL service was launched to the public on July 4, 1996, and was called Hotmail. It was one of the first web-based email services, allowing users to access their email accounts from any computer with an internet connection. Hotmail gained popularity rapidly and attracted millions of users, eventually leading to its acquisition by Microsoft in December 1997. It was later rebranded as Outlook.com.

The first email **alias** service was launched by RocketMail in 1997. RocketMail was one of the earliest web-based email providers allowing users to create several email aliases. RocketMail was later acquired by Yahoo! in 1997, and the service was integrated into Yahoo! Mail (which was and is still free).

The first email **alias** service that provided **unlimited** aliases was launched by Anonymizer in 1997. Anonymizer used a subscription-based pricing model to charge for their services. Users would typically pay a monthly or yearly fee to access Anonymizer's various features, including creating unlimited email aliases. This subscription model allowed Anonymizer to generate revenue and sustain its services' ongoing development and maintenance.

The service offerings just mentioned differed in one important way: email was free, but you had to pay for (unlimited) email alias support.

As spam became a problem, people resisted changing their email addresses because it was too troublesome, and the benefits of doing so were not compelling. After all, if you went to the trouble of changing your email address to avoid spam, not long after making the change, you would begin getting spam at the new address, so why bother?

Things have changed. It is now worth the bother. Let me explain.

Earlier Methods of Spam Control

SHORT OF CHANGING EMAIL addresses, the only way to control spam in the past was to learn to write and maintain personal spam filters (email rules) to handle those situations where the filters provided by ESPs were insufficient. The option to create your own filters has been available for a long time in most email applications, whether on the server side or a local system. You must have a personal computer to create local rules because this feature is unavailable for mobile mail apps.

Creating simple mail rules using guidance and templates is relatively easy for anyone motivated to do so. However, lacking incentive, motivation, and guidance, most people helplessly shrug at the ever-increasing pollution in their inboxes. Now, however, there is a better way.

Apple's new offering of support for unlimited aliases, packaged as a bonus feature when users upgrade from iCloud to iCloud Plus to add more storage for photos, essentially makes unlimited aliases available to all users at no extra charge. The "Hide My Email" (HME) feature is built directly into Safari, Mail, and iCloud settings. It arrived as part of the 2021 software update to iOS 15, iPadOS 15, MacOS Monterey, and iCloud.com. This and other bundled new features make Apple unique among major email service providers in providing a means of reducing the hidden dangers of spam email.

The alias handling feature has been perceived by most users as just a means of creating a few burner email addresses. A burner email is one that has a limited duration of perhaps just a few days. They are indeed handy for scoring a discount or downloading something for free. However, Apple probably didn't envision users using this feature to create a hundred or more aliases and use them <u>permanently</u>. Whatever the motive, it is now a moot point. The service has been launched with no limits.

I will use the diagram in Figure 2-3 to demonstrate how things could have been different had you used email aliases in the past. The diagram shows some of the largest data breaches from 2011 to 2021. There are thousands of data breaches yearly, but most are much smaller than those shown in the diagram. Many go unreported.

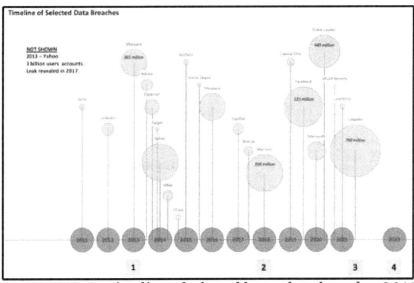

FIGURE 2-3. Timeline of selected large data breaches 2011 - 2021

1	Let's image that you had an account at the Myspace social network, which was hacked in 2013 exposing the email addresses, among other data, of 365 million users.
2	Let's image that you had an account at Marriott hotels which was hacked in 2018 exposing the email addresses, among other data, of 500 million users.
3	Let's image that you had an account at the Linkedin social network, which was hacked in 2021 exposing the email addresses, among other data, of 700 million users.
4	Had you been using the same email address from 2013 to 2023, that address would have remained on distribution lists used to send malicious spam for the entire period.

FIGURE 2-4. Breach timeline chart descriptions

In the scenario just described, had you been using email aliases unique to each account, you could have turned off the affected alias as soon as you learned of a breach in each case and then assigned a new one. As a result, malicious spammers would have lacked a valid address to send you spam in the first place. This is far better than relying on spam filters applied after an email has already been received.

As of the end of September 2023, corporations had reported 2,116 data compromises thus far for the year, according to the Identity Theft Resource Center (ITRC). That's already higher than the previous annual record of 1,862, set in 2021. The third quarter alone saw 733 total reported compromises, affecting over 66 million individuals. What's even more worrisome is that the actual number of breaches and victims is likely much higher than the ITRC's data shows. Officials at the ITRC note that transparency about attacks continues to get worse. And data breach notices, when filed, often lack details about how companies were compromised and victim details. James Lee, ITRC's COO, said "We also have new, clear evidence that companies are simply making a decision to not report a breach."

An important point is that no one knows which organizations will get hacked next month, the following month, etc. This solution offers the most powerful control over malicious spam there is.

Why You Should Act Now

CYBER RISK IS INCREASING, Let's look at some things that are changing as we enter the age of AI:

Changes in spam:

- The volume of malicious spam continues to climb.

- The increasingly evasive capability of malicious spam reduces the effectiveness of ESP spam filters.

- The effectiveness of malicious spam is growing as bad actors adopt AI tools.

- This improved effectiveness has several consequences: Near-perfect phishing emails will likely evade mail filters; the success rate of social engineering attempts will rise; the number of successful data breaches will increase, and more email addresses will be exposed.

Changes in email:

- The benefits of using aliases have not been wide-ly covered in the media and remain unclear to most users.

- The cost of unlimited email aliases has been re-duced or eliminated, removing a barrier to adop-tion.

- Creating and maintaining aliases have been sim-plified, removing a second barrier to adoption.

To Summarize, Now is the time to take action to gain control over spam before it gains control over you.

Chapter Three

PLANNING AND PREPARATION

A New Way Forward

THIS SOLUTION MAY SOUND a bit radical, but before you overreact, let me describe the process of implementation and the resulting benefits. A three-sentence summary of the process is as follows:

1. Get a new email address for personal business but don't share it with anyone.

2. Create aliases to protect it.

3. Update everywhere you have registered your former email address by changing it to an alias.

An alias has the same format as an email address so that it can be delivered from one domain to another on the Internet. In that sense, it is a valid email address. The receiving domain processes an alias differently, however. The receiving domain uses pattern matching to differentiate between a regular email address and an alias address. A normal email, after some filtering, is put directly into a single predetermined user inbox. An alias is subject to special processing to determine which inbox it will be sent to. For example, the target inbox of an alias can be changed without affecting the alias address itself, thereby providing flexibility. An alias can be turned off, reactivated, or discontinued at its owner's whim. An ordinary email address, in contrast, cannot be disabled and later reactivated. It can only be discontinued. The owner cannot recover it.

A normal email alias has no **time** limits on its use or validity. This is different from a "burner address", which is just an alias with a time limit placed on its use. A normal alias has the potential for long-term use, just like an email address. However, it is up to you, the "owner" to decide the permanency period.

How Alias Processing is Different

WHEN AN EMAIL IS sent to an alias, several changes typically occur before it is placed in a user's inbox by the alias processing service. These changes may vary depending on the specific email system or service being used, but here are some common things that happen:

1. **Address rewriting:** The alias processing service

typically rewrites the recipient address of the email. The recipient's original address, the alias, is replaced with the recipient's real email address.

2. **Header modification:** The email headers, including the From, To, and Subject fields, may be modified to reflect the original sender, the rewritten recipient, and any additional information or tags specific to the alias processing service.

3. **Spam filtering:** The alias processing service may perform spam filtering or apply specific filtering rules defined for the alias. This ensures that the email meets certain criteria and is not flagged as spam before it reaches the user's inbox.

4. **Forwarding or redirecting:** Depending on the alias configuration, the email may be forwarded or redirected to one or more email addresses. The alias processing service handles this forwarding process to ensure that the email reaches the desired destination(s).

5. **Allowlisting or blocklisting:** The alias processing service may apply any "allowlisting" or "blocklisting" rules set up for the alias. This can determine whether the email is allowed or blocked based on specific criteria, such as sender address, domain, or content.

6. **Virus scanning:** To protect the user's inbox from malicious content, the alias processing service may scan any email attachments or the entire email body. Appropriate actions are taken to mitigate the risk of viruses or malware being detected.

7. **Delivery to user's inbox:** Once the necessary modifications, filtering, and checks have been completed, the email is delivered to the user's inbox.

If you intend to keep the alias active until a certain event occurs, it can be said to be an "event-dependent address" - to distinguish it from a "burner address." Both are aliases, as I have just described.

I will use "ED address" to describe the event-dependent nature of aliases created to control spam in the two scenarios I outline below. If an ED address remains hidden from others and is not abused by spammers, it will be kept active. If it begins to attract spam, it will be disabled and replaced with another ED address.

Now let's examine which scenario might apply to you. It might be helpful to refer to (Figure 3-1) as you read these scenarios.

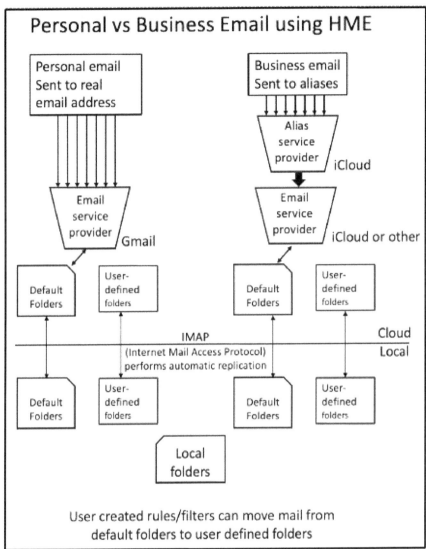

FIGURE 3-1. Splitting personal and business email

Scenario 1 – You have a single email address for all correspondence.

1. In this scenario, you have used the same email address for all correspondence with individuals or businesses.

2. You will want to split the incoming mail into two inboxes, one for personal and one for personal business (not employment-related email). You will choose an email service provider and open two new email accounts.

3. You will inform your correspondents of your new personal email address and ask them to begin using it instead of the old one, which you (tell them) you will be discontinuing. You will not be assigning email aliases to personal correspondents because they are an unlikely source of email addresses used by spammers and because doing so imposes some constraints on how you correspond in that you need to always REPLY to emails from individuals rather than create a NEW email to them. This issue will be discussed later.

4. You will create ED addresses (as previously described), which will route to the new email ad-

dress you will use for business. You will never share this new email address with anyone. Instead, you will share ED addresses.

5. With ED addresses, you log in and update your online account profiles to reflect how the account should communicate with you. Although it is not a requirement, I highly recommend assigning a unique ED address to each account. The reason for this will become clear shortly, but consider that updating your accounts to a new email address is no more difficult or time-consuming than changing to an ED address. Switching to a unique ED address for each account also adds no time to the steps involved. Preparation makes any process go more smoothly, and this task is no different. The preparation steps will be presented in the following sections.

6. Changing from an email address to an ED address will give you powerful control over spam once you have completed this process.

7. Note that making these changes in and of itself will not stop spam. It is the discontinuation of the old email address that will stop it.

8. Once you have completed these steps, any new spam sent to an ED address can be stopped cold by turning off the ED address and assigning it a new one. You will be able to prevent spam in minutes. Receipt of spam to an ED address is an

indication that the account has suffered a breach. You may even learn about it before the account knows about it. You will not have to wait to receive notice of the breach – if the account even bothers to inform users (many don't). You will already have dealt with it.

Scenario 2 – You have at least two real email addresses, one which you have used for only personal correspondence and the other for business.

1. In this scenario, you only need to create one new email address to replace the one you used for business correspondence.

2. You will perform steps 5 through 8 the same as in Scenario 1

The time and effort required to complete this process, using the guidance and templates provided in this book, largely depends on the number of online accounts you need to update with your new address. A good number for planning purposes is about 8 minutes per online account – if you are well organized when you perform the steps in the process.

You can do the math for your situation. For example, if you have about 100 online accounts and newsletter subscriptions, that would be 800 minutes or 14 hours to complete implementation. The most important point is that you only need to do this initial setup once.

Once you have these changes in place, you will only need a few minutes to update a single account when a breach occurs at one of the websites where you have an account. The setup process is the pain that's going to end the pain.

Making this strategic move puts you in control of spam in a way you never were before. It enables you to stop spam very quickly once you learn of the compromise of an ED address.

Gathering and Organizing Information

This book provides guidance, a template, and examples to implement spam control and to facilitate copying login credentials into (or updating existing ones in) a password manager. You should record information as you gather it, ideally using a tool such as a spreadsheet or a word processor. The advantage of using a digital method of recording is that you will later be able to copy/paste the information rather than having to re-type it.

Once you have gathered the information about your online accounts (the domain names) in this layout:

1. You will be prepared to create and label email aliases,

2. You will be prepared to update your login accounts

3. You will be prepared to create or update login records in your password manager.

4. You will be able to use it to quickly create email folders (either on the server or on your computer).

Selecting An Alias Service Provider

THERE ARE SEVERAL ALTERNATIVE offerings for creating and maintaining email aliases. Not all of those meet our needs which, briefly, are the following:

1. The service must offer unlimited or the ability to create up to 300 aliases at a reasonable cost.

2. The service must be an established company with strong financials that is likely to stay in business.

3. The service must provide both alias and reverse alias support.

4. The service must be easy to use.

The top three offerings in the market that meet our needs for creating email aliases are Apple (`https://support.apple.com/en-us/HT210425`), Annonaddy (rebranded as Addy) (`https://addy.io/`), and Proton's Simple Login (`https://simplelogin.io/`).

There is also a new service (3Q2023) launching in the US (only) called Cloaked (`https://www.cloaked.app/`) based on the principle of hiding personal contact information. Cloaked is just coming out of beta and for a monthly fee, offers not only unlimited email aliases but also "cloaked" phone numbers, and a password manager. Time will tell if it is successful in gaining the trust of new users.

In this document, I will explain how to implement email aliases using Apple's "Hide My Email" (HME) feature available to those who subscribe to iCloud Plus.

To use the HME feature, you must first have an Apple account, and to create such an account, you must have an Apple device or access to one for at least a short period because you cannot create an Apple iCloud account unless you are doing so using an Apple device (Computer, mobile phone, or tablet). Many people have an Apple iPhone and an account on Apple's iCloud. The HME service is an additional benefit to any Apple user who has paid to upgrade storage above the "free" tier. Since so many people have already done this because of the need for more space to store photos, HME is essentially already included in the subscription price for the extra storage.

Introducing Apple Hide My Email

ALONGSIDE IOS 15, APPLE debuted a new iCloud+ service, which adds additional features to all paid iCloud accounts. The iCloud+ service includes several new features, and the one we will explore in detail here is called "**Hide My Email,** "or HME. With HME, iPhone, iPad, and Mac, users can create unique, random email addresses forwarded to a personal inbox. For example, if you need to sign up for an online account, you can use a random Apple-created email alias.

All the emails sent to the alias are forwarded to your real email address so you can respond if needed, but the sender does not see your real email address. Should you begin receiving spam emails, you can deactivate or delete the email alias that is the problem to put a stop to it.

You can create separate aliases for each website because Apple does not limit how many aliases you can create.

You can choose the real email address that your HME addresses forward to. By default, it will select your Apple ID. Still, if you have other email addresses associated with your account (which can be done under Settings > Apple ID > Name, Phone Numbers, Email), you can select another email address option.

HME Email works for both receiving and sending emails. If you respond to an incoming email forwarded to an HME address in the Mail app, Apple will continue to obscure your email address in the reply.

You can also create an HME Email address using Safari or an app when signing up for an online account. The "Hide My Email" option will come up as a suggestion, and if you tap it, Apple will offer a randomly generated email address for you to use and email you to confirm its creation. This is why when you first open HME, you may find a few aliases you didn't explicitly create. Another situation when HME will create an alias on your behalf is when you use "Sign in with Apple" to log in to a website. Apple does not share your real email address with third parties and will create an alias for the third-party website. This is one more example of how Apple respects and protects your privacy.

As you create aliases, I recommend you record each one in a spreadsheet, assigning one to each online account domain you have listed. You should use the label field in HME to give the alias the label of the domain name so that you can easily keep track of it. Ideally, you will create a unique alias for each domain. Although this is optional, not doing so somewhat defeats the purpose of this method of spam control.

Let's look at the template for information gathering (See Figure 3-2).

Online account domain name (from URL)	Description	Email Alias assigned	Change completed
1password.com	Password manager		
aa.com	Airline		
adobe.com	Software provider		
amazon.com	Store		
chase.com	Bank		
citi.com	Bank		
consumercellular.com	Cellular service provider		
costco.com	Store		
dropbox.com	App		
equifax.com	Credit bureau		
ghin.com	Golf association		
google.com	Search, mail, etc.		
hilton.com	Hotel		
linkedin.com	Social media		
nationalgeographic	Media		
nest.com	Thermostat		
netflix.com	Media		
questargas.com	Utility company		
zoom.com	Video conferencing service		

FIGURE 3-2 - Template for domain-to-alias mapping for online accounts and subscriptions

Understanding the data collection form:

- Column 1 of the template contains the domain name from the website URL where you have an account. You should consider using the domain name as the label for an alias as you create them in HME

- Column 2 is for entering a description of the domain and is optional.

- Column 3 is for entering the new email alias you have created for this account. This information will later be used when updating your online account profile.

- Column 4 indicates you have completed updating the account to use the new email address.

You may wish to add additional column in your spreadsheet to record additional information For example, you could add a column to indicate that you have also changed the password to an account to a stronger one. Using a password manager in and of itself doesn't improve security. Change your passwords to stronger ones and insuring that they are not reused is what improves security.

Using Hide My Email to Create Email Aliases

TO GENERATE A RANDOM email alias using iCloud+, follow these steps:

1. Open System Settings.

2. Tap your Apple ID name, then tap iCloud.

3. Tap Hide My Email.

4. On the lower left corner of the window Tap Plus (+) to create New Address (an alias).

5. Label it – Enter a descriptive label for the new email alias in the Label Your Address text field. Since the email alias will communicate with a domain, using the domain name as the label makes sense. If you want, add a note to help you remember what the email is for, why you created it, and any other helpful information. If you're unhappy with the random email alias, tap Use Different Address to generate a new one – but you will be given only a few options before the app cycles back to the one originally offered. Finally, tap Next to complete the process and you will see this final confirmation. Tap Continue to complete the process. You will be returned to the list of all of the aliases you have created.

Hide My Email

You can add a note to help you remember how you use this email address.

↻ contour_leading0e@icloud.com

LABEL | aa.com

NOTE | American Airlines website

Cancel | Continue

FIGURE 3-3 HME - Labeling your alias

Your new email alias and any others you've created through iCloud+ (or "Sign in With Apple") will appear in the list. Note there is a search bar at the top of the window to find aliases you have previously created by address or label or by contents of the Note field.

Take note of the HME email address pattern, which will likely contain **"0*@icloud.com"** in the email address where the * can be any alphabet letter. An example of this is shown in the screenshot.When you create more than one alias at a time with HME, you may encounter a small problem. You might create ten or so aliases and then receive an error message saying you cannot proceed further. Apple throttles alias creation so you may need to return a few minutes later to continue to create more aliases. This may happen repeatedly but eventually you can create as many aliases as you require.

In any case, you want to estimate how many aliases you need, and this is based on how many accounts you have that you will assign them to. Start with about 100. You can always create more at any time. Think of this as creating a jar of aliases, each unique, of course, for use when you begin the implementation steps discussed in Chapter 5: Alias Implementation.

It is always good to have a few unassigned aliases in case you must provide your email address to a store or a person you meet. I create half a dozen extras which I label as Unassigned 1, Unassigned 2, etc., in HME. I then copy these to my Notes app, where I can reference them quickly and document how I use them. Later, I update the labels in HME. I also have a few extras printed in an easy-to-read size and monospaced font on a business card-sized piece of paper that I keep in my wallet. This is useful when I want to hand it to someone to copy from. This simplifies the information exchange. Again, I update the labels in HME soon afterward.

Chapter Four

MANAGING AND USING EMAIL ALIASES

BEFORE MOVING ON TO implementing aliases using HME, we should first review how easy it is to manage them.

You can view and manage all your randomly generated email aliases in the Settings app.

You can temporarily deactivate aliases or delete them entirely. Once you've **deleted** an alias, note that you can no longer reactivate it or receive messages sent to it. Any emails sent to the alias following deactivation or deletion will be flagged as an invalid email address and returned as undeliverable to the sender. This is a good thing.

Before you disable or delete an alias, you should first manually change the alias for any accounts, newsletters, or other services affected if you wish to continue receiving emails from those sources.

Deactivating an alias

At any time, you can deactivate an email alias by following these steps:

1. In the Hide My Email section of the Settings app, tap on the email address you'd like to deactivate.

2. Then, tap Deactivate Email Address.

3. Finally, confirm that you'd like to deactivate it.

The email alias remains available for you to reactivate and use anytime. You will only receive messages sent to it once you do so.

Reactivating an email alias

Here's how to reactivate an email alias:

1. In the Hide My Email section of the Settings app, tap on Inactive Addresses.

2. Then, choose the address you'd like to reactivate from the list.

3. Tap Reactivate address and confirm that you'd like to reactivate the email address.

Deleting an email alias

You'll only see the option to delete an email address once you've deactivated it. Here's how to delete an inactive email alias:

1. In the **Hide My Email** section of the Settings app, tap on **Inactive Addresses**.

2. Then, choose the inactive email address you'd like to delete permanently.

3. Tap **Delete Address** and confirm that you'd like to delete the address permanently.

Using Aliases for Personal Business Communications

USING AN EMAIL ALIAS for business communications isn't a problem because you usually call a business or go to its website to <u>initiate</u> communications. Most emails sent from companies to customers are largely one-way communications. Businesses send invoices, shipping notices, and advertising. They don't often expect a reply. Many emails from organizations often use the "Reply To" address which routes responses to an unmonitored inbox named "noreply."

Businesses also use aliases with a "Reply To" so that replies are routed to a different designated address instead of sending the message back to the originator. For example, a business might send invoices to customers from billing@bigcompany.com. Most customers will pay the bill and not reply, but those with an issue perhaps get routed to the customer support department. In such a situation, a "Reply To" address will invite customers to respond directly to a support email address (e.g., customersupport@bigcompany.com.) Similarly, when sending a shipping notice to a customer, a reply might be routed to the package tracking department, and so on.

Email service providers (ESPs) often identify reply-to addresses in email headers as an indication that the sender is a legitimate source—in other words, not spam. A phishing email attempting to impersonate a large business that does NOT have a reply to address will likely not make it through spam ESP spam filters.

An alias service makes several changes to the email header information before placing an email in your inbox. One change is to modify the "Reply" or "Reply To" information so that if you do respond, the email first must be sent back through the alias processing service, where a "reverse alias" action is performed. This special handling will replace the original alias address with the "From" or "Reply-To" field. This means that when the recipient receives your reply, it appears to come from the alias address, thereby hiding your real email address.

Because of such changes, it is easy to tell when the HME service handled an email.

Using aliases for personal correspondence

Let's say you have a friend named Bill with whom you exchange the occasional email. You have decided to ask Bill to begin sending emails to an alias address instead of a real email address. You create an alias for Bill to use that send him an email asking him to update his contact information to accommodate your request.

You also ask him to send you an email using the new alias address, explaining that you want to confirm that he has the right address.

You should save at least one email from Bill that was sent to your alias so you can use it and subsequent ones for your replies.

This will become a habit change for you but not for Bill. If you want to email Bill, you must open his email folder and reply to an email he sent using the alias you assigned him. Doing so will cause it to be sent back through the alias service for proper handling, including hiding your real email address.

If, instead, you compose a new email to Bill, that new email will not go through the alias service, and as a result, it will arrive in his inbox from your real email address, causing some confusion.

Unfortunately, you cannot just change Bill's contact address to the one that would route a new email through the alias service. There are other differences between the header information in an email you reply to and one you compose as new that causes this "trick" to fail. Your email will be returned with a "delivery not authorized" error.

This isn't a deal breaker but requires a habit change. It doesn't impact Bill once he has changed his address for you in his contact book.

Mail Privacy Protection: An Additional Feature of iCloud+

MARKETING EMAILS, NEWSLETTERS, AND some email clients use an invisible tracking pixel in email messages to check to see whether you've opened an email. This approach to determining whether an email address is valid and actively being used is also exploited by hackers. Hackers often clean up email address information after obtaining the data from hacking a company or other organization. They don't want to waste sending phishing emails to accounts that are no longer valid or not being used, but more likely, they don't want to suffer a bad reputation among their criminal community for selling or renting to others a list of emails that are not valid and active.

Apple enables users to stop this practice with **Mail Privacy Protection**, included as part of iOS15 and its MacOS counterpart.

Before this and still the situation with other email service providers, you had to turn off HTML formatting for mail to avoid the impact of "tracking pixels" in email.

FIGURE 4-1. Mail privacy protection

Many email apps provide an option to block remote images, which effectively prevents tracking pixels from working, but Mail Privacy Protection is an easier-to-use, universal solution. <u>It is NOT on by default and needs to be enabled</u> in the Mail section of the Settings app.

Mail Privacy Protection prevents email senders from tracking whether you opened an email, how many times you viewed an email, and whether you forwarded the email. It does not block remote images but instead downloads all remote images in the background regardless of whether you've opened an email, essentially ruining the data.

It also has the added benefit of hiding your IP address so senders cannot determine your location or link your email habits to your other online activity.

Apple routes all content downloaded by the Mail app through multiple proxy servers to strip your IP address, and then it assigns a random IP address corresponding to the general region you're in. Email senders see generic information rather than specific information about you.

Mail Privacy Protection is an alternative to blocking all remote content, and if the feature is enabled, it overrides the "Block All Remote Content" and the "Hide IP Address" settings.

Chapter Five

ALIAS IMPLEMENTATION

THE SPREADSHEET YOU CREATED in Chapter 3 can but need not be retained after your work is done. That is because the information will be available in other locations accessible to you (specifically through the email alias app) and your various online accounts and even documented in the logon records in your password manager.

Since the information is not very sensitive, it is not a danger to keep it around for handy reference. You will probably find it useful to track new additions over time as you create new online accounts or discontinue existing ones; in this case, you have a handy reference for aliases you might want to reassign rather than create new ones. When you make such changes, remember to update the labels (domain names) you assigned to the aliases in your alias management application (HME in our example).

Updating Online Accounts

Once you have the list of accounts and aliases, you can update your online accounts to communicate with you using the email aliases instead of your real email address. This process involves logging in to each account through your web browser and updating your account profile information. In most cases, updating your email address will likely confirm the ownership of the new address (which will then be an alias) by receiving and responding to emails sent to you.

This process will take, on average, between 5 to 10 minutes. Therefore, if you have 100 accounts, you can see that this will take several days to accomplish from start to finish. However, note that you can do this over weeks rather than all at once. Spend a little time each day and soon you will be finished. Understand that using aliases puts you in control for the future but does not limit the amount of spam you receive that is sent to the real email address you will be replacing. You must close the original email account to eliminate spam sent to your former address.

You will likely miss a few accounts as you go through the replacement process, so as you receive mail from those you overlooked, you can handle them by following the same steps as before. I suggest waiting a few months before closing an email account completely. One way to ensure you don't miss something is to set up mail forwarding from the email account to an alias. You can create a special folder and a rule to automatically move mail to that folder. The rule would compare the "TO" address to the alias. This way, any remaining emails sent to the original email address would be placed in this folder, alerting you that some accounts were overlooked as you assigned aliases.

The goal is to assign a unique email alias to each account. Using an alias does not control spam in and of itself. What it does is put you in control when spam "happens". If there is a data breach, you can easily deal with the situation by deactivating the alias you used for the impacted account and assigning it a new one. The original alias, likely exposed in the breach, will no longer be valuable to the hackers who breached the account. Any spam they send to the old alias will be "bounced" back as an invalid address.

As you access your online accounts to make these changes, you will likely want to update the "user-id" stored in your password manager (assuming you are using one). In my implementation, I found that about 80% of accounts use the email address as the userid for login. Your new user ID will become the unique email alias you have created instead of (formerly) your real email address.

While You're At It (updating accounts, that is)

WHEN YOU LOG IN to an account to change your email address to an alias, why not take care of a few other matters?

The data collection template (Figure 3-2) could be amended to include a column to indicate that you have also updated your password. Perhaps you were using one that needed to be strengthened. If the account allows it, I recommend you make your passwords at least 15 characters long. Also, you may have reused a password or two. This would be a good opportunity to sort that out by making them all unique. If you decide to change an account password, you may have to respond to an email or, in some other way, verify that you are the account owner and that it is you making the change, just as you did when changing your userid from email address to an alias you have chosen for that account.

Most password managers provide a password generation tool to help you select a new password, but you may want to use a standalone tool for this purpose as you may find more flexibility and ease of use by doing so. My favorite such tool, available for free from the App Store, is PSWD by developer Kerem Erkan. Figure 5-1, shown below, is a screenshot of this tool, demonstrating its extensive flexibility in creating passwords and ease of use.

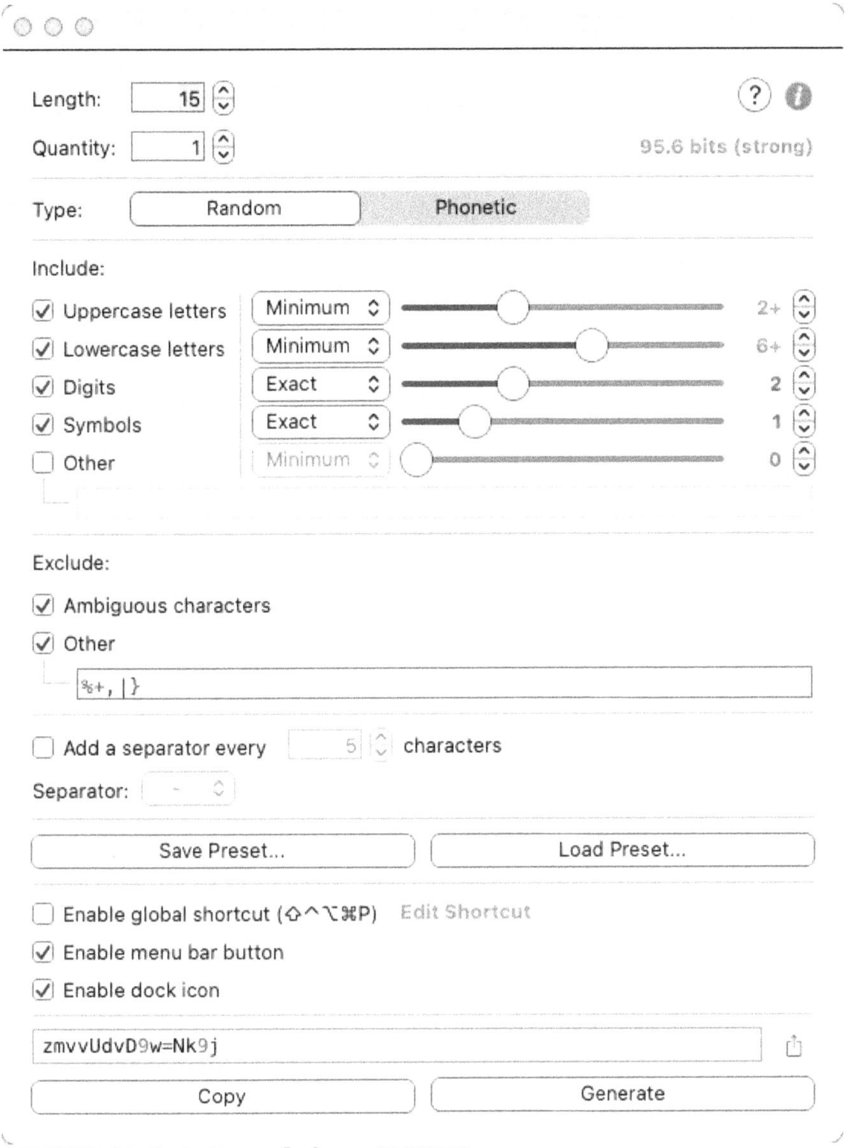

Length: 15 ⌄   ? ⓘ

Quantity: 1 ⌄   95.6 bits (strong)

Type: [Random]   Phonetic

Include:

☑ Uppercase letters   [Minimum ⌄] ━━━━━━━◯━━━━━━━━ 2+ ⌄
☑ Lowercase letters   [Minimum ⌄] ━━━━━━━━━━◯━━━━ 6+ ⌄
☑ Digits   [Exact   ⌄] ━━━━━━◯━━━━━━━━━ 2 ⌄
☑ Symbols   [Exact   ⌄] ━━━◯━━━━━━━━━━━━ 1 ⌄
☐ Other   [Minimum ⌄] ◯━━━━━━━━━━━━━━ 0 ⌄

Exclude:

☑ Ambiguous characters
☑ Other

    %+,|}

☐ Add a separator every   5 ⌄ characters
Separator: [- ⌄]

[Save Preset...   ] [Load Preset...   ]

☐ Enable global shortcut (⇧⌃⌥⌘P)   Edit Shortcut
☑ Enable menu bar button
☑ Enable dock icon

zmvvUdvD9w=Nk9j   ⬆

[Copy   ] [Generate   ]

FIGURE 5-1. Standalone PSWD app to create strong passwords

Finally, while logged in to an account, this might also
be a good time to review the communications settings,
determining what kind and how many emails you receive. This will be discussed in more detail in Chapter 7.
Many companies send out several categories of emails,

allowing you to select which ones you want to receive and opt out of those you don't. If you do not make your selections explicitly known, the default is often to subscribe you to everything.

One important setting usually found under most on-line accounts' privacy and security section is sharing your information with others, such as affiliates, business partners, etc. If you see such an option, choose to opt-out of sharing, and you will further reduce unsolicited emails.

Overview of The Complete Process

To BEGIN THE WORK of implementation, it is handy to open five apps on your desktop:

1. Your alias management app – Apples HME in our example.

2. Your spreadsheet, where you created a list of your online accounts (domains)s you assign(ed) each one. You will use this to record the email alias you assign to each domain and the completion of this process.

3. Your browser, where you will log in and update online accounts.

4. Your email app, where you will respond to emails

sent by domains to verify ownership of the alias you assign each.

    a. Open your email client to be ready to respond to change verifications you will get from the account where you will change the existing login credentials to use a new user, which will be the email alias created in step one.

    b. Be aware that email verifications may arrive in your junk or spam folder.

5. Your password management app, where you will create (or update) login records to automate the login process.

    a. The process steps listed below will create a new record for a login.

    b. If you already have a record for an account, you will be updating that record.

Process Steps

WITH THE APPS OPEN on your desktop, here is a sequence of steps to perform. This sequence assumes you have not implemented a password manager but will be doing

so as part of implementing email aliases. It demonstrates the creation of login records in a password manager (1Password in this example) as part of updating
online accounts to communicate with you using aliases.

1. Create a new alias with HME.

  a. ->System Settings -> Appleid -> iCloud ->
     Hide My Email.

  b. Click "+" to add a new alias.

  c. In the "label" field, type in the Domain name
     to which you will assign the alias.

  d. Click the CONTINUE button, then click the
     COPY LABEL button to capture the alias name
     to the system clipboard.

2. Go to the Alias-to-Domain mapping spreadsheet
   window (that you have created in preparation for
   this process).

  a. Go to the row for the Domain you will be
     updating.

  b. Paste the new alias in Column D of the spread-
     sheet row (Command+V).

3. Go to your browser window.

  a. Open a new tab.

  b. Type in a Domain (such as "Michaels.com")
     and press return.

    c. Go to the "Sign in" page.

    d. Copy the URL from the browser address bar (Command+C).

4. Go to your password manager window.

    a. Click on the "New Item" button.

    b. Choose "Login" as the new item record type.

    c. Go to the URL field in the new record.

    d. Paste the Login URL (Command+V).

    e. Go to the "Username" field.

    f. Type your Username (typically your email address such as "xxx@gmail.com").

    g. Go to the "Password" field.

    h. Type in your current login password for the domain.

    i. Go to the Login name field.

    j. Type in "Login – domain name."

    k. Click SAVE.

5. Return to the spreadsheet window.

    a. Go to the row for the domain.

    b. Copy the alias from column D (Mac Command+C)

6. Return to the browser window.

    a. Log into Account.

    b. Look for your account "Profile."

    c. Look for CHANGE email address.

    d. Paste in the alias (Mac Command+V) to re-place the email address information.

    e. Save the change (usually a Save button).

    f. The Domain may want to confirm that you own the New Alias address by emailing you.

7. Go to the email window and respond to confirm ownership of the alias.

8. Return to the Password Manager window.

    a. Go to the "Username" field.

    b. Update the Username field by pasting the alias (Mac Command+V).

    c. Save the change using the Save button.

9. Return to the Domain-Alias mapping spread-sheet window.

    a. Type in "COMPLETED" in Column E.

    b. Save the updated spreadsheet.

You will repeat the above steps for each domain/alias pair.

Chapter Six

AUTOMATING MAIL HANDLING

YOU CAN AUTOMATE MAIL handling tasks such as moving mail to folders, deleting mail based on various criteria, forwarding mail, etc. Mail automation is done by creating filters or rules, as they are sometimes called, depending on the app. Creating rules is optional but beneficial and therefore recommended.

Rules should be written after email aliases are implemented because email header information gets modified when processed by whatever alias routing service you use. Because of the changes, a previously created email rule may not work as originally written.

Creating Mail Folders

While each account is given a unique email alias, when it comes time to create mail folders to store correspondence, in some cases, you may prefer to store emails from multiple accounts in a single folder. For example, you may have accounts with three airlines, but you will create one folder called airlines for emails from any of them.

In most cases, however, you will likely want to create one folder per account. Because your mail app sorts mail folders alphabetically, you would end up with a group of mail folders that looks very much like the list of domain names used to label your aliases in Figure 3-1.

The first decision to make regarding creating mail folders is whether you want your folders to be on a mail server or just on a local machine. When I say local machine, I mean a computer such as a Mac or Windows system. You cannot create local (only) mail folders on a mobile device. If you want your mail stored in folders available on any device, you should make your folders on the server. If you choose to have folders in both locations, you should create two lists using the template in Figure 3-1. One list will be for IMAP folders that you create on the server through webmail and the second will be local folders you create using your email client.

Mail Folders On the Server

THESE INSTRUCTIONS DEMONSTRATE HOW to create what are called "IMAP" folders using iCloud.com. If you are

using another ESP such as Gmail, you would log into
Gmail through your web browser.

- Log in to your Apple account on iCloud.com
  through your browser. Go to iCloud mail. The
  panel on the left-hand side of your screen will
  display default Mail folders: Inbox, VIP, Drafts,
  Sent, Junk, Trash, and Archive. You can create additional folders and subfolders to organize your
  email by clicking on the plus sign to the right,
  where you see the word folders at the bottom
  of the panel. Enter a name in the box and press
  return. A new folder will be created and opened
  (empty) in the right-hand panel. Most ESPs offer
  some variation of the standard folders just mentioned.

- You can also create a subfolder by selecting the
  folder where you want to add a subfolder, clicking the plus sign, typing the new subfolder's
  name, then pressing Return. You can also click an
  existing folder and drag it onto another folder.
  To move a subfolder to the top level, click the
  subfolder, then drag it onto the folder header in
  the Mailboxes list.

When you create additional folders, they will become
available on any device with mail sync turned on in
iCloud settings. When you create folders on a (mail)
server, assuming you are using what is called IMAP
(not the less common POP) support, all folders will be
automatically synchronized to all devices that access

mail on the server - just as is the case with your inbox (and other ESP provided default folders).

Other email apps may offer some options here, such as allowing you to select which folders will be synchronized. The spam or junk folder is an example of a default folder on a server you can choose to replicate to a local device (or not).

Of particular significance is that you will have to create your rules on the server if a rule involves moving mail to folders (as opposed to flagging, forwarding, or deleting them) them. If you have multiple email accounts, you will have to create folders and rules on each account, whereas if your folders are on a local machine, you can sort mail from multiple accounts into a single set of folders.

Mail Folders On a Local Device

FOR ANY FOLDERS YOU create on a local device, be aware that you must create mail rules on that device (in the email client app) for moving mail to folders that (only) exist there. Mail filtering rules available on the client side are generally more robust than those on the server side, so there might be cases where you want or need to use the advanced features that cause you to choose local folders rather than server-based ones. Another consideration is that server-based folders replicated to other devices will take up space on those devices. This may be a problem in some cases where storage space is limited.

I prefer local mail folders; all my mail handling rules are created on my local machine. All the mail I move from my inbox to local folders will not be available on any of my other devices – which is an intentional choice. There are several ways to get mail to my phone when I want it there – such as when I am on vacation. Generally, I prefer to handle mail exclusively on my computer rather than on my mobile devices. I use messaging apps to communicate on my mobile devices.

You can use a Windows or Apple computer to create local folders. You can also choose the mail (client) app you use on your computer. Creating folders and rules (filters) is independent of alias processing.

It is important to set up your mail client app to access the inbox provided by iCloud, and you can take it from there. These instructions are rather generic. My local machine is a Mac, and I use the Apple Mail client app.

Organizing Mail Folders

SOME PEOPLE PREFER INDIVIDUAL folders rather than a folder and sub-folder hierarchy. Others may have just a few folders and use the search function to find what they want. My personal preference is to have a folder hierarchy in which I have folder groups and, under these, subfolders with one for each domain. This allows me to organize folders such that all banks are together, all stores together, and so on. I find it easier to locate an account I am looking for when they are grouped this way. This is accomplished by creating additional folders with no contents but establishing a folder hierarchy like how your computer file system looks. Whatever works for you is fine. It is easy to change by renaming folders and moving them between groups.

The diagram in Figure 3-1 showed how you might use one email address for personal correspondence and aliases for business email. This would require two real email addresses. All mail sent to aliases is sent to a single real email address.

I have created a folder hierarchy that works for me, and you should make one that works for you. The table below describes the folders I have created in the mail client on a local machine. All my folders are in my LO-CAL mail folder hierarchy. I do not presently use any user-defined IMAP folders. If I had any IMAP folders, they would be defined on the server through webmail, but, of course, they would also appear on my local machine because of the replication feature provided by IMAP mail handling.

These folder groupings (categories of accounts) are ones I have defined to meet my own needs. Figures 6-1a and 6-1b below describes how I use them to store emails I receive in my inbox. You can manually sort emails, but it is more productive to automate this process using mail rules which I have done and will describe later.

Each top-level folder is preceded in this table with one or more underscore characters (_) to show the level in the folder hierarchy for purposes of this description. The mail app automatically indents folders, so my actual folder names do not include these characters in the names. Also, note that I have given names to some of these folders that begin with a number. This is to force the sort order how I want it to appear when I open my mail app. This allows me to reduce scrolling when handling my mail.

One issue with creating groups is that sometimes an account may fall into two or more groups. An example is Amazon.com which is a place to buy stuff, but Amazon.com is also a provider of media streaming services. I decide which group to put it in and be done with it.

LOCAL FOLDERS	
_1 – Most Frequent Correspondents	If, like me, you decide to create a unique mail folder for most accounts, you might want to create a group such as this where you can move mail folders that you frequently access to the "top of the stack" to reduce scrolling.
_2 – Potential Spam	If you implement sorting with mail rules (recommended) then you will want a place to put mail (using rules) that didn't get handled after all existing rules have been processed. When an email is placed in this folder, it may be that my rules are incomplete and need to be updated, or the item is spam or even malicious spam from an unknown sender.
_3 – Newsletters	Used for newsletter subscriptions. There will likely be cases where you subscribe to a news or information source by do not have to create an online account.
_4 – Personal Email	Used for folders from individual correspondents such as family and friends.
_5 – Business Email	Used for correspondence related to personal business of myself and my family. Note that all the following groups have two underscore characters at the front of the name to indicate these mail folders are below this one in my hierarchy of mail folders.
__Apps-for-General-Use	Used for accounts representing apps or information resources on the internet that I use. Examples include folders for Google, Dropbox, etc.
__E-Commerce	Used for retail websites.
__Financial-Accounts	Used for banks, stockbrokers, credit providers, credit bureaus, sources of income, insurance, and credential management tools. Not all the folders

FIGURE 6-1a. Examples of mail folder names

	are strictly financial but are important and likely have financial implications.
___Financial-Accounts-Inactive	Used for accounts that perhaps been closed to maintain a record of correspondence.
___Government	Used for all government service websites other than those that provide income (see financial accounts).
___Healthcare	Used for all healthcare related websites.
___Local Services	Used for locally provided services such as utilities, pest control, lawn care, etc.
___Media	Used for all media accounts such as video and music streaming services as well as news sources whether on-line or in print format.
___Memberships	Used for accounts with membership organizations such as the US golf association, Lions club, AARP, etc.
___Products	This group contains folders for accounts related to physical products that I own. These are important for warranty registration, and to be notified of software updates where appropriate.
___Social Media	This group contains folders related to social media apps.
___Travel	This group contains all folders related to travel, lodging, transportation, dining out, etc.
YY - Email from uncommon domains	When you receive an email from any unusual TLD, there is a higher chance that the communication is from a malicious spammer and warrants increased scrutiny. The YY prefix causes this folder to appear near the bottom of my list of folders.
ZZ - Archives	Any mail I want to ensure I hold on to and don't mistakenly delete I put here.

FIGURE 6-1b. Examples of mail folder names (continued)

Creating Mail Rules

EMAIL RULES CAN BE used to automate and streamline your email management process. The specific use cases will vary depending on your individual needs and preferences. Most email clients and services provide options to create rules or filters, allowing you to customize your email management according to your personal needs.

Here are some common uses for handling personal email: Moving mail from your inbox to a mail folder, Spam Filtering, Email Forwarding, Categorization, and Tagging.

1. **Email Inbox and Folder Organization:** Email rules can be used to automatically sort incoming messages into specific folders based on criteria such as sender, subject, keywords, or recipient. For example, you can create a rule to move emails from various senders into dedicated folders, making finding and managing relevant emails easier. After you have created aliases, a better approach which I have adopted is to look at the TO address. Based on the alias the mail was sent to corresponds to a domain and therefore a mail folder since I named my folders according to domains.

2. **Spam Filtering:** Email rules can assist in filtering out spam or unwanted messages. You can create rules to automatically move emails from known spam senders with certain keywords or specific patterns into the spam or junk folder. This feature will be largely unnecessary if you begin using aliases to keep unwanted mail out of your inbox.

3. **Email Forwarding:** If you need to receive copies of specific emails at multiple addresses or share them with family members, email rules can automatically forward incoming messages to designated email accounts.

4. **Categorization and Tagging:** Email rules can assist in categorizing or tagging emails based on specific criteria. This can help you organize and filter emails for easier retrieval and reference later or to indicate some conditions, such as emails that include an attachment or even an attachment of certain types. Rules can be used to change text colors or add a colored flag that has a meaning you define yourself.

Some Useful Rules

MAIL RULES ARE SPECIFIC to your implementation. It would be great if you could import rules developed by someone else, but since the online accounts you have and the folder names you create in your email account are unique, this isn't feasible. However, I can provide a few examples that you can use to create your rules.

The most common rules are those that move mail from your inbox to a folder you have created for correspondence with a specific sender, such as one of your banks. Let's take an example:

1. Assume that you have an account at Chase Bank. The correct URL for this bank is https://www.ch ase.com/.

2. You create an email folder called "chase.com" or "chase" based on the domain name.

3. A rule might look like the following statement:

If [choice 1=ANY] **of the following conditions are met:** [choice 2=FROM] [choice 3=ENDS WITH] <u>chase. com,</u> **Perform the following actions:** [choice 4=MOVE MESSAGE] **to the mailbox:** <u>chase.com</u>.

The mail rule as it would appear in Apple mail settings is shown in Figure 6-2 below:

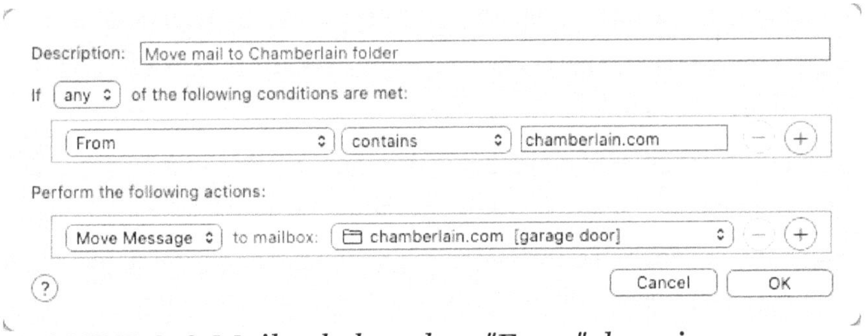

FIGURE 6-2 Mail rule based on "From" domain

Looking at options available in creating a rule, the first is the If statement, where you can choose "ANY" or "ALL."

This continues with specifying what it is about the email you want to examine for comparison purposes. There are about 20 such options; in this case, the choice

of "FROM" means that we will look at that informa-
tion/attribute of the email. In other words, we will be
examining the sender's address information.

The next step is providing further clarification regarding what it is about the information we want to
examine. In this example, we have chosen to check if
the from address ends with a specific character string.
There is a handful of other choices, such as "CON-
TAINS" or "IS EQUAL TO."

Finally, the rule specifies what (character string) we
will compare the address to. In the example, we have in-
dicated the character string "chase.com." All mail is sent
to some domain (a server) somewhere on the Internet,
so email addresses always end with a domain name.

If the conditions are true, the next step is to indicate
the desired action.

There are about a dozen possible actions on offer, and
in our example, we have chosen "MOVE MESSAGE" to
a mailbox (a mail folder).

Note that to the right of the IF statement, there are
two buttons, a Minus and a Plus. The Plus button allows you to add more than one condition. If you have
multiple specified conditions, the Minus button will
be active, allowing you to remove a previously created
condition. Similarly, buttons to the right of the action
statements will enable you to add more than one action.
For example, you can change the color of the text in one
action (statement) and then move the mail to a folder
in the next action statement.

As you can imagine, this can get quite involved. That
is why most people do not create rules. If you keep

it simple such as moving flagging messages, it isn't so difficult to do as it might seem.

Another thing to note is that you need to create folders to hold mail before you make rules because when creating rules in Apple Mail, you select an already existing folder in the rule creation window. There is no option to **enter** a mailbox (folder) name.

The domain name of the sender will change

When you use aliases, the sending domain of the email that arrives in your inbox will no longer be the original one. It will be the domain of the alias service instead. The email rule just described would no longer work as intended because it would look for chase.com as the sender's domain. If you are using Apple's HME alias service, the email you receive will be sent from the icloud.com domain. You would then need to modify the rule to work as you originally intended.

But wait ... there is a better way to write rules after you have assigned unique aliases to each account. The rule just mentioned examines the "FROM" address. If you look at any email sent to you using an alias, it will look like the image in Figure 6-3 below.

Note that at the top of the email, there is a clear indication that the email was forwarded to you by HME. Next, notice that when you click on the "TO:" address field, you will see the alias you had assigned to this sender's domain. (It has been defaced in the image).

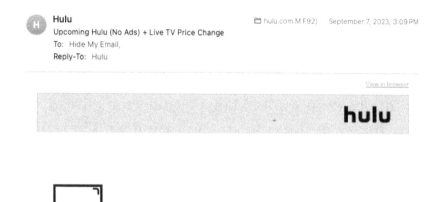

Upcoming Hulu (No Ads) + Live TV Price Change

FIGURE 6-3 Mail received via HME

You can create a rule that looks at the "TO" address, as depicted in Figure 6-4 below. In this example, if the "TO" address matches the alias you assigned, it will be moved to the folder for mail from this domain. (It has been defaced in the image).

FIGURE 6-4 Mail Rule based on "TO" address (now an alias)

This is very useful. Let me explain. Only you know what aliases you created or how you assigned them. If one of your aliases gets compromised and spammers begin using it, they have no idea you associated it with a specific domain. Let's say you have an alias assigned to an account (website) for gym you belong to. Let's further say that you have created a mail rule that moves mail sent to this alias to your gym folder.

Now, suppose a spammer sends a malicious email to that alias purporting to be UPS and suggesting there is a problem with package delivery. In that case, the email will end up in your gym folder because of the matching "TO" the alias name. When you open the gym folder, you will see the email from the spammer alongside legitimate emails (from your gym). The spam email will be so obviously out of place that you will recognize that it is bogus (see photo in Figure 6-5 below).

Obviously, the alias assigned by you to the gym account has been compromised, so after marking the fake email as spam and deleting it; you will want to change the alias assigned to the gym account and update your mail rule to use the new alias for comparison. Problem solved. No more spam.

FIGURE 6-5. There is an obvious impostor in this mail folder

The order of rules is important

There are some rules that I place at the top of my list. Let's look at the purposes of some of those.

A rule that alerts me when an email contains an attachment that could be dangerous. The first mail rule tags an email with a red flag if it has an executable file attachment – which can be highly dangerous (Figure 6-6). I wouldn't want to open one of those from ANY-ONE without first checking it out. Most likely I'm going to delete it.

FIGURE 6-6. Sample rules for flagging EXECUTABLE attachments

A second attachment-related rule sets a diffferent flag color for files with potentially dangerous filetype such as zip files, office files, and PDFs (Figure 6-7).

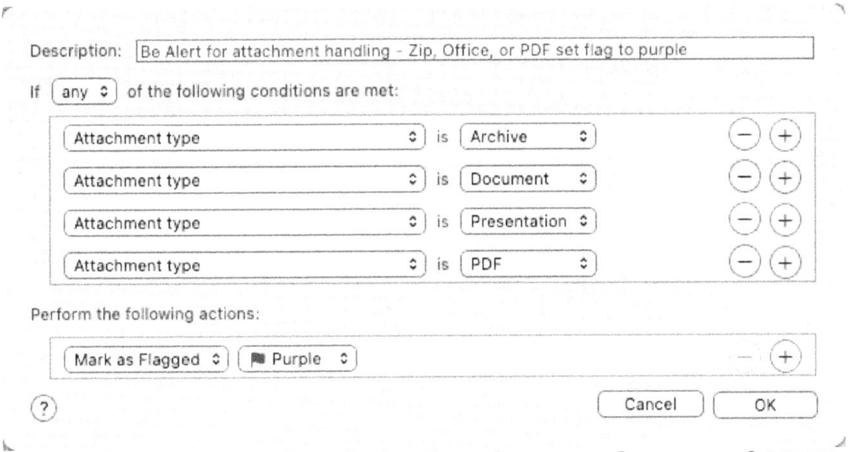

FIGURE 6-7. Sample rule for flagging attachments of OTH-ER types

Here is a rule that flags files that contain the word "Unsubscribe" in the body of the email (Figure 6-8). If you are a cautious person, you will verify the link behind an unsubscribe request is going to go where you think it should and not just accept that it is safe and go there.

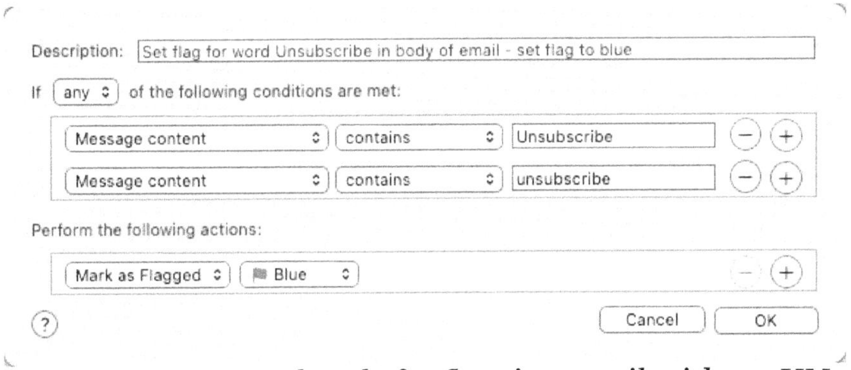

FIGURE 6-8. Sample rule for flagging email with an UN-SUBSCRIBE link in the body

There are also some rules that I place at the bottom of my list:

A rule that identifies emails from unusual domains potentially abused by malicious spammers (Figure 6-9). This will catch mail from domains in Russia, China, Brazil, etc. without having to make a more complex rule that looks for .RU, .CN and so forth.

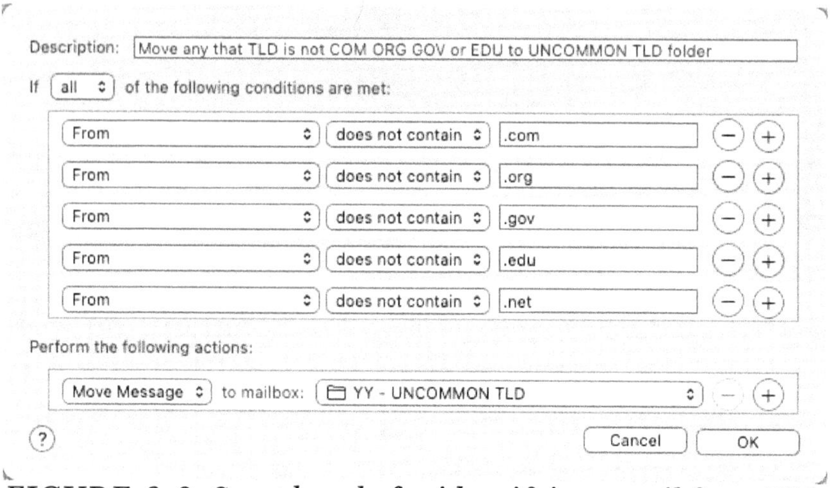

FIGURE 6-9. Sample rule for identifying email from UN-USUAL domains

A rule that handles mail for senders that are not in my contacts and that I have not previously moved to another folder (Figure 6-10).

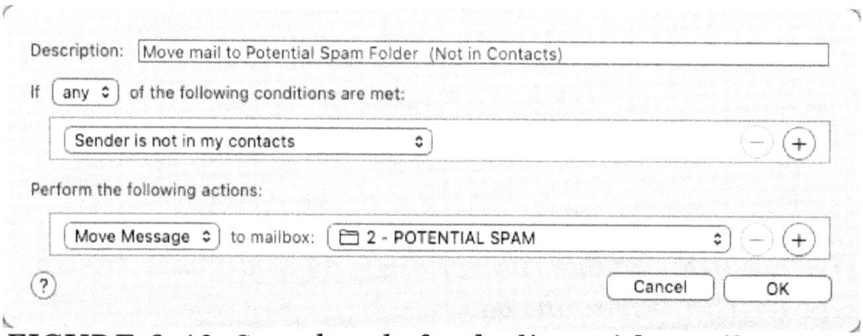

FIGURE 6-10. Sample rule for dealing with email not already handled

Some, including myself, have written rules to identify email that is likely junk based on specific words in a message's subject line or body. Here are some commonly used words or phrases that have been associated with spam: "Urgent, "Important," "Exclusive," "Limited time," "Free," "Offer," "Discount," "Guaranteed," "Act now," "Cash," "Win," "Won," "Congratulations," "Special promotion," "Save," "Unclaimed," "Inheritance," "Lottery," "Opportunity," "Investment," "Money-back guarantee."

With the approaches identified in this book, you should be able to control spam <u>without</u> resorting to this never-ending craziness. The rule-writing process is now reduced to mostly sorting your mail because you have control of your address like never before.

Once proper rules are created, spammers cannot easily fool you by sending mail that does not come from the correct domain. If sent from a legitimate domain, your rules only move mail to your designated folders. Otherwise, they will stay in your inbox or be moved (by you with a rule) to the "**Potential Spam**" folder I previously mentioned that you might want to create. Goodbye, spoofed email and malicious spam!

Chapter Seven

MANAGING BENIGN SPAM

ORGANIZATIONS SEND EMAILS THAT generally can be classified as relational or transactional. Each serves a different purpose and is used in specific scenarios. Let's explore the differences between relational and transactional emails:

Relational Emails: Relational emails primarily focus on building and nurturing customer relationships. They are typically sent as a part of ongoing communication efforts and aim to engage and connect with recipients on a more personal level. Advertising makes up the bulk of relational emails. Here are some key characteristics of relational emails:

1. **Content:** Relational emails often contain personalized content, such as welcome messages, newsletters, updates, announcements such as new product launches or special pricing, blog subscriptions, event invitations, or customer surveys.

2. **Recipient base:** Relational emails are usually sent to an entire customer or subscriber base, targeting a broader audience.

3. **Frequency:** They can be sent regularly, such as weekly or monthly, to maintain consistent communication and keep recipients informed and engaged.

4. **Design and branding:** Relational emails often reflect the business's branding, featuring a visually appealing design, company logo, and consistent messaging to reinforce brand recognition.

Businesses, especially larger ones, often enable consumers to opt out of relational emails. If you log in to an account, you can search for communications settings in your account profile and select which communications you wish to continue receiving and which you don't. This is the way you reduce benign spam most effectively.

Some businesses, especially small and medium-sized businesses (SMBs), share contact information with others they refer to as affiliates or business partners. When updating your account communications profile, you should look for this setting and disallow it so that you don't start getting emails from companies you have never heard of or done business with.

Transactional Emails: Transactional emails are triggered by specific actions or events and are typically related to a customer's interaction with a business. They provide essential information or confirmations and are

considered transaction related. Here are some characteristics of transactional emails:

1. Purpose: Transactional emails are designed to
   provide information or confirmations related to
   specific customer actions, such as purchases, order confirmations, shipping notifications, password resets, account updates, or appointment
   reminders.

2. Timing: They are usually triggered immediately
   after the customer's action, delivering timely and
   relevant information.

3. Personalization: Transactional emails can include personalized details, such as order details,
   tracking numbers, or account-specific information.

4. Compliance: Transactional emails often comply
   with legal requirements, including opt-out or
   unsubscribe links, order details, or privacy policy
   information.

5. Minimal marketing content: Transactional
   emails may contain subtle cross-selling or up-
   selling opportunities, but they primarily focus
   on delivering transaction-related information
   rather than promotional content.

A consideration in developing mail rules that sort
mail based on the account is that many businesses,
large and small, use third-party email marketing plat-

forms. An example of a popular email marketing platform is Mailchimp which allows companies to create, send, and manage email campaigns to their subscribers.

You should not be surprised if you receive transactional emails from one domain and relational emails from the same account but from another domain. Writing rules to account for this situation is very easily accomplished. Still, it is something you need to recognize is happening, and nothing unusual about it should raise a red flag. However, you should also know that bad actors may send malicious emails that mimic relational or transactional mail types.

The best way to reduce spam from legitimate sources is to unsubscribe. Note that clicking on a link in an advertising email often removes you from a single list, and most companies have more than one. Therefore, the best way is to log in to the account and look in your profile for something like **Communications Preferences.**

When updating your preferences, at some accounts, such as banks, you can select what is sent to you and choose the communications channel that will be used. You may want to receive a notice via text message to your phone about credit card charges above a certain amount but have all other communications sent only to email.

In most cases, however, you will choose which types of communication you want to receive and which you don't. If you only want to receive transactional emails but eliminate relational (advertising) emails, this is the best way.

Often when you register for a new account or sign up for a newsletter, you are asked to agree to receive communications. You can often deselect the opt-in checkbox when you sign up, and you won't have to return to the account later to turn off the spam.

Figure 7-1 below shows an example of communications preferences in the account section of a website:

Profile & Preferences

Personal Info   Security   Communications

Sign up for text messages from Lowe's.

We'll send a code to this number to make sure it's you. After verifying your number, choose the communications you'd like to receive below.

Begin Verification

How would you like Lowe's to connect with you?

You can change your preferences or unsubscribe at any time.

Email	SMS Text Messaging	Subscription	Description
Off / On			
		News & Ideas	New products, special offers and ideas for the whole family.
		Deal of the Day	An amazing deal you won't want to pass up sent straight to your inbox every day.
		Recommended for You	Product and promotional updates you might be interested in.

FIGURE 7-1. Example of COMMUNICATIONS PREFERENCES

Controlling Spam Email From Third Parties

SOME BUSINESSES HAVE A checkbox to indicate your agreement to share information with affiliate or business partners. If you overlook this, you will likely receive emails from organizations you never heard of or do business with. Emails you receive from third parties may or may NOT offer the opportunity to "unsubscribe." Once they have your email address, there is no way for you to take it back. With an alias, however, you could disable it to stop them from sending you spam.

Another way you get emails from third parties is when a company purchases mailing lists from data brokers. They often do this to reach potential new customers. The data broker industry is a largely hidden but very big business. There are hundreds of players in this space.

Only by removing your email address from circulation can you avoid communications from third parties that have obtained your email address. This is why it is important, when you begin anew, to keep your real email address private. You will confuse data brokers if you implement unique aliases for each account. Over time, they will likely collect a few of your aliases, but you will be in charge. You can shut down any aliases that attract spam and replace them with fresh ones.

Chapter Eight

IMPLICATIONS OF USING EMAIL ALIASES

THERE IS ONE IMPORTANT consideration of implementing unique aliases for each of your online accounts: It will become quite impossible to continue to avoid using a password manager to manage your credentials.

The reason is that there will be many more things to "remember" than just passwords.

Most people use the same email address for each account. As a result the only thing that protects access is the chosen password. Because so many people choose weak passwords and reuse them, many websites have adopted an additional means of protecting accounts.

One such method, with two implementations, is called two-step verification (or two-factor authentica-tion). This means of security requires an additional step in the authentication process. One method involves an account sending a text message to an account holder with a one-time numeric code that must be input into a web form to continue. A second, safer approach requires the user to obtain the one-time code using an additional app called an authenticator.

To make matters worse, some websites require the use a specific authenticator app. There are quite a few of these available, so you could end up with several installed on your mobile device, adding even more confusion.

Using a unique alias and a unique password adds an additional layer of security. It is like having a safe deposit box with two keys, but different in that you have both keys (see Figure 8-1). One is a bit more public, and the other kept private.

FIGURE 8-1. Safe deposit box with two keys

Look upon this as a positive outcome. It will force many people to get off the fence and finally adopt the use of password manager. Various sources indicate that only about one in five (about 45 million) US internet users currently use a password manager. Nearly two in three people still track their passwords by memorization or hand-written notes. One study reports that web users not using a password manager were found to be three times more likely to experience identity theft than those who properly use them.

Perhaps many simply need to be made more aware of the security and productivity benefits they are missing out on by not using a password manager. A password manager stores the user ID and password in an encrypted vault and automates the login process. You don't have to remember so many things. That is the point. Let's take a moment to review the many benefits offered by this time-saving application:

1. **Convenience:** Password management apps simplify the process of managing credentials required for authentication. They allow you to generate complex, unique passwords for each online account and store them securely. This eliminates the need to remember or manually enter passwords, making it more convenient to access your accounts across different devices.

2. **Saving time:** With a password manager, you don't have to spend time looking for or recovering forgotten passwords. The app securely stores your credentials and automatically fills them in when needed, saving you time and frustration. You will avoid repeated password resets.

3. **Improved accessibility:** Many password managers offer synchronization across multiple devices. This means your passwords are securely

stored and accessible on your computers, smartphones, and tablets. You can access your passwords from anywhere, making it convenient to log in to your accounts on different devices.

4. **Enhanced password strength:** Password managers can generate strong, complex passwords for you. Strong passwords are typically lengthy, randomized, and difficult to crack. You significantly improve your online security by using unique, robust passwords for each account.

5. **Secure password sharing:** Password managers often provide certain sharing functionality. This allows you to safely share passwords with trusted individuals without compromising security.

6. **Simplified password change management:** When you update a password for an account, the password manager can automatically store the new password and update it across your devices, eliminating the need for manual updates.

7. **Secure form filling:** Besides passwords, many password managers offer form-filling capabilities. They can securely store and auto-fill personal information, such as names, addresses, and credit card details, at the touch of a button, making online form submissions more efficient and secure.

8. **Encrypted storage:** Password managers use ro-

bust encryption techniques to store your passwords securely in a digital vault. Encryption
ensures that your sensitive information remains protected even if the password manager's data is compromised. This vault is suitable
for storing digital copies of important personal
and family documents such as driver's licens-
es, passports, birth and marriage certificates,
property titles, proof of military service, social
security cards, and even heirloom photos.

9. **Family protection:** In case of incapacitation
or untimely death, trusted family members
can access your information using features designed for this purpose.

By providing convenience, time-saving features,
and synchronization across devices, password managers streamline the management of important and
sensitive information for more than just your online
accounts.

A password manager is an essential tool for digital
security, and there are quite a few solid password
management apps available to choose from, both
free and for a reasonable fee. Just do a Google search
using the term "Password manager" and find one that
best fits your needs.

Most password manager vendors offer training
videos, and 1Password is no exception. You can visit the
company support page at https://support.1password.
com/explore/get-started/ to see what 1Password offers.

There are also many videos created by third parties on services such as YouTube.com.

How Using Single Sign-On May Affect The Use of Aliases

THERE IS A RELATIVELY new method of account login (the authentication process) offered on some websites. In the industry, this method is referred to as single sign-on (SSO).

For many users drowning in website logins and constantly using **"Forgot My Password"** prompts to get into random accounts, a "**Log in With Google**," "**Log in With Facebook**," or "**Log in With Apple**" button can look a lot like a lifeline. The services provide a quick way to continue whatever you're doing without having to set up an account and choose a new password to guard it. But while these "**Single Sign-On**" tools are convenient and offer some security benefits, they're not the panacea you might think.

SSO may be a gift from heaven for security, but it could be a mistake for protecting privacy.

Apple launched a new single sign-on (SSO) tool in June 2019 in response to similar initiatives by Google, Facebook, and others. As described onstage by Apple software chief Craig Federighi, users would encounter the service as a simple sign-in button, presented as an alternative to setting up a persistent username and password (credentials) for a given service. But where Google and Facebook use those buttons to link you to your broader advertising profile, gaining a wealth of ad-tracking data in the process. Apple's service is designed to give the minimum necessary data. So, if you use an SSO solution, you should consider using the Apple offering if given a choice – for privacy reasons.

Whereas others share information between apps, Apple's approach doesn't even share your email address with an account you use the service with, instead directing each app to a different redirect email address operated by Apple. With a different redirect for each app, it is far more difficult for third parties to correlate information by comparing emails. And when a user wants to cut ties with an app, breaking the redirect will sever the connection entirely.

Problems with Single Sign On

The SSO schemes offered by big tech companies have some obvious advantages. For example, they're developed and maintained by companies with the resources to bake in strong security features. Take "Sign in With Apple," which lets you use TouchID or FaceID to log into any number of sites.

For all its convenience, consumer SSO has some real drawbacks; the first is that doing so creates a single point of failure if something goes wrong.

The security of any online account will be only as strong as the password used to protect it unless you add

Two-factor authentication. If your password or access token gets stolen from an account you use for SSO, all the other sites where it was used could be exposed. And not only do you have to trust the companies that offer SSO to protect your privacy and security, but you also must trust all the third-party websites offering these options to implement them correctly.

The inherent risks aren't just hypothetical. In September 2018, Facebook disclosed a massive data breach that impacted at least 50 million of its users and, among other things, exposed any other account data those people logged into using Facebook SSO. Facebook invalidated the access tokens when it detected the breach (we don't know how long it took to detect it). Still, the incident underscored the potential ripple effects of any consumer SSO breach.

Many consumer SSO schemes also present practical issues with account recovery. Determining who is responsible for helping you troubleshoot problems can take time and effort. For the average user, there are a daunting number of factors in the choice of committing to a password manager versus using SSO.

If you don't have the time or energy to worry about the nuances, though, much less managing different passwords in different ways, a password manager is a proven, mature, one-stop solution that's always helpful—whether a certain site offers SSO or not.

Even if you are using SSO, you should attempt to establish the use of an email alias in place of a real email address. Due to implementation differences, this might not be possible to implement if you choose an SSO solution to log in to your accounts. After login to an account using SSO, you should examine your account profile to see if you still have the option of changing an email address to an alias. Due to various implementations, this may not be possible. For example, if you sign in with Google, you may have to stick with your Gmail address, but this remains unclear. Suppose you cannot change to an alias. In that case, you should consider disconnecting from Google and switching to an alternative such as Apple, if that is an option, or switching to the legacy login approach to create a login using an alias address and a password.

It's All Up To You Now

As we all know, reading about something and doing it are two different things. Know that cyber criminals are highly motived to make money by stealing it from you in some form or fashion. It is up to you to get equally motivated to take action to defend yourself.

Hopefully you are convinced that the method of cybercrime prevention presented in this book is worth your time and effort. Also, hopefully, I have presented the information to you in a way that is thorough, understandable, and actionable.

Be safe and free of those spammy "flies and mosquitos" in your inbox.

QR Codes

These QR codes are provided to facilitate easy access to online resources.

To check and see if your email address or phone number have been exposed on the Dark web in a data breach, visit the website HaveIBeenPwned.com

To learn more about personal cybersecurity best practices, visit the author's website at ScamSmarts.com

ABOUT THE AUTHOR

Jon has two degrees in Computing Science, is a US Air Force veteran and spent most of his career working at IBM.

In retirement he spends some of his time each day doing research into the field of cybersecurity.

In both Fall and Spring he teaches courses on cybersecurity in the community education program at Utah Tech University in St. George, Utah.

He is delighted to have identified how to use a decades old technology, email aliases, in a new way to make email safe again.

ACKNOWLEDGEMENTS

Special thanks to Dr. Rick Campbell for providing the inspiration to write this book and for his assistance in documenting the process of implementation.

www.ingramcontent.com/pod-product-compliance
Lightning Source LLC
La Vergne TN
LVHW051245050326
832903LV00028B/2579